INCREASING THE AGS POLARIZATION

Related Titles from AIP Conference Proceedings

675 SPIN2002: 15th International Spin Physics Symposium and Workshop on Polarized Electron Sources and Polarimeters
Edited by Y. Makdisi, A. Luccio, and W. MacKay, June 2003, 0-7354-0136-5

631 New States of Matter in Hadronic Interactions: Pan-American Advanced Study Institute
Edited by H.-Thomas Elze, Erasmo M. Ferreira, Takeshi Kodama, Jean Letessier, Johann Rafelski, and Robert L. Thews, October 2002, 0-7354-0086-5

570 SPIN 2000; 14th International Spin Physics Symposium
Edited by Kichiji Hatanaka, Takashi Nakano, Kenichi Imai, and Hiroyasu Ejiri, June 2001, 0-7354-0008-3

549 Intersections of Particle and Nuclear Physics: 7th Conference, CIPANP2000
Edited by Zohreh Parsa and William J. Marciano, December 2000, 1-56396-978-5

512 Nuclear Physics at Storage Rings: Fourth International Conference: STORI99
Edited by Hans-Otto Meyer and Peter Schwandt, June 2000, 1-56396-928-9

482 RHIC Physics and Beyond: Kay Kay Gee Day
Edited by Berndt Müller and Robert Pisarski, July 1999, 1-56396-878-9

343 High Energy Spin Physics: Eleventh International Symposium
Edited by Kenneth J. Heller and Sandra L. Smith, 1995, 1-56396-374-4

To learn more about these titles, or the AIP Conference Proceedings Series, please visit the webpage **http://proceedings.aip.org/proceedings**

INCREASING THE AGS POLARIZATION

Ann Arbor, Michigan   6-9 November 2002

EDITORS
A. D. Krisch
A. M. T. Lin
University of Michigan
Ann Arbor, Michigan

T. Roser
Brookhaven National Laboratory
Upton, New York

SPONSORING ORGANIZATIONS
University of Michigan
Brookhaven National Laboratory

Melville, New York, 2003
AIP CONFERENCE PROCEEDINGS ■ VOLUME 667

Editors:

A. D. Krisch
A. M. T. Lin

Spin Physics Center
University of Michigan
Ann Arbor, MI 48109-1120
USA

E-mail: krisch@umich.edu
alilin@umich.edu

T. Roser
Building 0911B
Brookhaven National Laboratory
Upton, NY 11973-5000
USA

E-mail: roser@bnl.gov

Authorization to photocopy items for internal or personal use, beyond the free copying permitted under the 1978 U.S. Copyright Law (see statement below), is granted by the American Institute of Physics for users registered with the Copyright Clearance Center (CCC) Transactional Reporting Service, provided that the base fee of $20.00 per copy is paid directly to CCC, 222 Rosewood Drive, Danvers, MA 01923. For those organizations that have been granted a photocopy license by CCC, a separate system of payment has been arranged. The fee code for users of the Transactional Reporting Service is: 0-7354-0130-6/03/$20.00.

© 2003 American Institute of Physics

Individual readers of this volume and nonprofit libraries, acting for them, are permitted to make fair use of the material in it, such as copying an article for use in teaching or research. Permission is granted to quote from this volume in scientific work with the customary acknowledgment of the source. To reprint a figure, table, or other excerpt requires the consent of one of the original authors and notification to AIP. Republication or systematic or multiple reproduction of any material in this volume is permitted only under license from AIP. Address inquiries to Office of Rights and Permissions, Suite 1NO1, 2 Huntington Quadrangle, Melville, N.Y. 11747-4502; phone: 516-576-2268; fax: 516-576-2450; e-mail: rights@aip.org.

L.C. Catalog Card No. 2003105096
ISBN 0-7354-0130-6
ISSN 0094-243X
Printed in the United States of America

CONTENTS

Preface vii
A Workshop Photo ix
Paper Not Included in this Volume xi

First Polarized Proton Collisions at RHIC 1
    T. Roser, L. Ahrens, J. Alessi, M. Bai, J. Beebe-Wang, J. M. Brennan,
    K. A. Brown, G. Bunce, P. Cameron, E. D. Courant, A. Drees, W. Fischer,
    R. Fliller, III, W. Glenn, H. Huang, A. U. Luccio, W. W. MacKay,
    Y. Makdisi, C. Montag, F. Pilat, V. Ptitsyn, T. Satogata, S. Tepikian,
    D. Trbojevic, N. Tsoupas, J. van Zeijts, A. Zelenski, K. Zeno,
    A. Deshpande, K. Kurita, K. Krueger, H. Spinka, D. Underwood,
    M. Syphers, I. Alekseev, D. Svirida, V. Ranjbar, J. Tojo, O. Jinnouchi,
    M. Okamura, and N. Saito

Thoughts and "Facts" from the AGS Polarized Proton Runs during the 1980's ... 9
    L. Ahrens

Overcoming Intrinsic and Coupling Spin Resonances in the AGS 15
    M. Bai, L. Ahrens, and T. Roser

Overcoming Depolarizing Resonances at COSY 30
    A. Lehrach, U. Bechstedt, J. Dietrich, R. Gebel, B. Lorentz, R. Maier,
    D. Prasuhn, A. Schnase, H. Schneider, R. Stassen, H. Stockhorst,
    and R. Tölle

20% Partial Siberian Snake in the AGS 40
    H. Huang, L. Ahrens, M. Bai, K. A. Brown, W. Glenn, A. U. Luccio,
    W. W. MacKay, C. Montag, V. Ptitsyn, T. Roser, N. Tsoupas, K. Zeno,
    V. Ranjbar, H. Spinka, and D. Underwood

Review of Polarized Proton Beam Acceleration at KEK-PS in the 1980's ... 50
    C. Ohmori, S. Hiramatsu, H. Sato, and T. Toyama

AGS Pulsed Quadrupoles: History and Future 58
    A. D. Krisch

OPPIS Upgrade for 2003 Polarized Run in RHIC 61
    A. Zelenski, J. Alessi, B. Briscoe, A. Kponou, S. Kokhanovski, V. Klenov,
    V. LoDestro, J. Ritter, and V. Zubets

AGS Lattice Changes to Eliminate Weak Intrinsic Resonances 67
    A. Lehrach and V. H. Ranjbar

The AGS CNI Polarimeter 77
    G. Bunce, I. G. Alekseev, A. Bravar, S. Dhawan, H. Huang, V. Hughes,
    G. Igo, O. Jinnouchi, V. Kanavets, K. Kurita, Z. Li, W. Lozowski,
    W. W. MacKay, Y. Makdisi, S. Rescia, T. Roser, D. N. Svirida, C. Whitten,
    and J. Wood

A No-Depolarization Theorem for Rotator-Aided Resonance Crossing 81
    D. W. Sivers

Spin Matching from AGS to RHIC 84
    W. W. MacKay and N. Tsoupas

Matching of Siberian Snakes 93
    G. H. Hoffstaetter
Workshop Highlights and Summary 103
    T. Roser

Participants List 109
Final Agenda 111
Author Index 113

PREFACE

There was a good reason for organizing the Workshop on *Increasing the AGS Polarization*. The surprising spin effects discovered at lower energy accelerators, make the new RHIC multi-hundred-GeV polarized proton collider especially important. As described in this Workshop's opening talk, the four Siberian snakes in the two RHIC rings seem capable of maintaining high polarization during acceleration and storage. RHIC's main problem was the low injected polarization from the AGS. This problem was exacerbated by the weak temporary AGS power supply, which slowed down the acceleration cycle and thus made the AGS depolarization stronger. On the positive side, this exacerbation highlighted that the AGS polarization had to be significantly improved.

The Organizing Committee, consisting of Thomas Roser and myself, felt that it was necessary to document what is needed for increasing the AGS polarization, so that it could be used, precisely and quickly, to make next year's polarized RHIC run even more successful. I believe that we succeeded in this goal due to the dedication, hard work, and creativity of the participants in the Workshop. I hope that the proceedings of this Workshop will help the Brookhaven staff in their goal.

I would like to thank Debbie Walls and Alison Fried of Michigan and Mary Campbell of Brookhaven who were responsible for the secretarial and administrative tasks at the Workshop. They all contributed to the success of the Workshop. I would also like to thank Dr. A. M. T. Lin, who did much more editing than both of the other editors. The main justification for Thomas Roser and I being listed as editors is our work in organizing the Workshop.

A. D. Krisch
Ann Arbor, March 2003

The following paper was included in the Workshop program,
but has not been submitted for publication:

Orbit Matching with AGS Helical Snake

Ernest D. Courant, Brookhaven National Laboratory, Upton, NY 11973 USA

Professor Courant gave a similar talk at the *15th International Spin Physics Symposium* held at Brookhaven National Laboratory in September 2002. Thus, he decided not to publish it again.

This paper is published in *SPIN 2002: 15th International Spin Physics Symposium and Workshop on Polarized Electron Sources and Polarimeters,* edited by Y. Makdisi et al., AIP Conference Proceedings 675, Melville, New York, 2003.

First Polarized Proton Collisions at RHIC[1]

T. Roser*, L. Ahrens*, J. Alessi*, M. Bai*, J. Beebe-Wang*, J. M. Brennan*,
K. A. Brown*, G. Bunce*[†], P. Cameron*, E. D. Courant*, A. Drees*,
W. Fischer*, R. Fliller III*, W. Glenn*, H. Huang*, A. U. Luccio*,
W. W. MacKay*, Y. Makdisi*, C. Montag*, F. Pilat*, V. Ptitsyn*,
T. Satogata*, S. Tepikian*, D. Trbojevic*, N. Tsoupas*, J. van Zeijts*,
A. Zelenski*, K. Zeno*, A. Deshpande[†], K. Kurita[†], K. Krueger**,
H. Spinka**, D. Underwood**, M. Syphers[‡], I. Alekseev[§], D. Svirida[§],
V. Ranjbar[¶], J. Tojo[‖], O. Jinnouchi[††], M. Okamura[††] and N. Saito[††‖]

*BNL, Upton, NY
[†]RIKEN BNL Research Center, Upton, NY
**ANL, Argonne, IL
[‡]FNAL, Batavia, IL
[§]ITEP, Moscow
[¶]Indiana Univ., Bloomington, IN
[‖]Kyoto Univ., Japan
[††]RIKEN, Japan

Abstract. We successfully injected polarized protons in both RHIC rings and maintained polarization during acceleration up to 100 GeV per ring using two Siberian snakes in each ring. Each snake consists of four helical superconducting dipoles which rotate the polarization by 180° about a horizontal axis. This is the first time that polarized protons have been accelerated to 100 GeV.

INTRODUCTION

Polarized proton colliders will open up the completely unique physics opportunities of studying spin effects in hadronic reactions at high-luminosity high-energy proton-proton collisions. It will allow study of the spin structure of the proton, in particular the degree of polarization of the gluons and antiquarks, and also verification of the many well-documented expectations of spin effects in perturbative QCD and parity violation in W and Z production.

The Brookhaven Relativistic Heavy Ion Collider (RHIC) is the first hadron accelerator and collider consisting of two independent rings. It is designed to operate at high collision luminosity over a wide range of beam energies and with particle species ranging from polarized protons to heavy ions. The RHIC center-of-mass energy range of 200 to 500 GeV[1] is ideal in the sense that it is high enough for perturbative QCD to be applicable and low enough so that the typical momentum fraction of the valence quarks is about 0.1 or larger. This guarantees significant levels of parton polarization.

---

[1] Work performed under the auspices of the U.S. Department of Energy

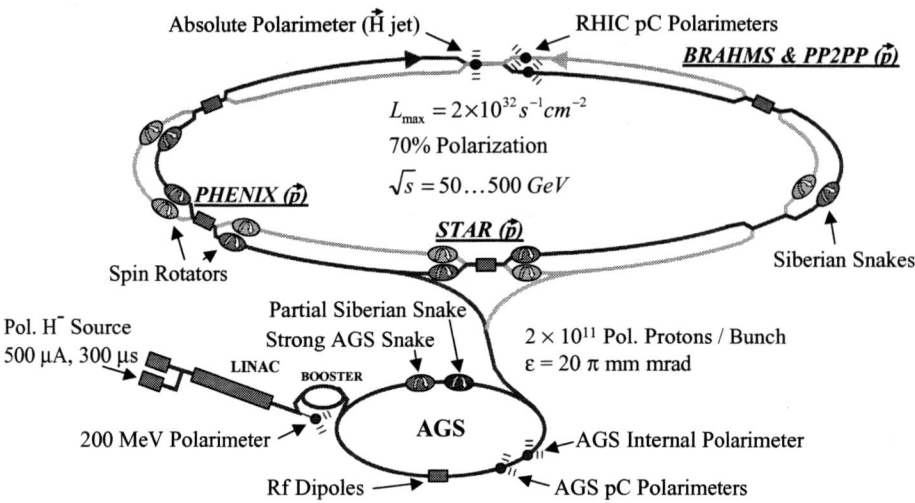

FIGURE 1. Layout and design parameters for the Brookhaven polarized proton collider. The eight spin rotators and the absolute polarimeter were not installed for this run. The beam parameters given correspond to the goal luminosity of 2×10^{32} cm^{-2} s^{-1}

During the second RHIC run polarized proton beams were successfully accelerated to 100 GeV and stored and collided with a peak luminosity of about 1.5×10^{30} cm^{-2} s^{-1}. With the two full Siberian Snakes installed in each of the two RHIC rings essentially all beam polarization was preserved during acceleration and beam storage in RHIC.

SPIN DYNAMICS, RESONANCES AND SIBERIAN SNAKES

Accelerating polarized beams requires the control of both the orbital motion and spin motion. The evolution of the spin direction of a beam of polarized protons in external magnetic fields, such as those existing in a circular accelerator, is governed by the Thomas-BMT equation [2],

$$\frac{d\vec{P}}{dt} = -\left(\frac{e}{\gamma m}\right)\left[G\gamma \vec{B_\perp} + (1+G)\vec{B_\parallel}\right] \times \vec{P}$$

where the polarization vector P is expressed in the frame that moves and rotates with the particle's velocity. This simple precession equation is very similar to the Lorentz force equation:

$$\frac{d\vec{v}}{dt} = -\left(\frac{e}{\gamma m}\right)\left[\vec{B_\perp}\right] \times \vec{v}.$$

Comparison of these two equations readily shows that, in a purely vertical field, the spin rotates $G\gamma$ times faster than the orbital motion. Here $G = 1.7928$ is the anomalous magnetic moment of the proton and $\gamma = E/m$. $G\gamma$ gives the number of full spin precessions for every revolution and is also called the spin tune ν_{sp}. At top RHIC energies ν_{sp} reaches about 450.

The acceleration of polarized beams in circular accelerators is complicated by the presence of numerous depolarizing spin resonances. During acceleration, a spin resonance is crossed whenever the spin precession frequency equals the frequency with which spin-perturbing magnetic fields are encountered. There are two main types of spin resonances corresponding to the possible sources of such fields: imperfection resonances, which are driven by magnet errors and misalignments, and intrinsic resonances, driven by the focusing fields. The strengths of both types of resonances increases with beam energy.

The resonance condition for imperfection depolarizing resonances arise when $\nu_{sp} = G\gamma = n$, where n is an integer. Imperfection resonances are therefore separated by only 523 MeV energy steps. The condition for intrinsic resonances is $\nu_{sp} = kP \pm \nu_y$, where k is an integer, ν_y is the vertical betatron tune and P is the superperiodicity.

With a localized spin rotator that rotates the spin by the angle δ about a horizontal axis the spin tune is given by

$$\cos(\pi \nu_{sp}) = \cos(\pi G\gamma)\cos(\delta/2).$$

The spin tune can never reach an integer for any non-zero δ and therefore all imperfection resonances are avoided. A 'full Siberian snake' [3], which is a 180° spin rotator, will make the spin tune a half-integer and energy independent. Therefore, neither imperfection nor intrinsic resonance conditions can ever be met. In the presence of strong resonances the spin rotation of the snake has to be much larger than the total spin rotation from the resonances.

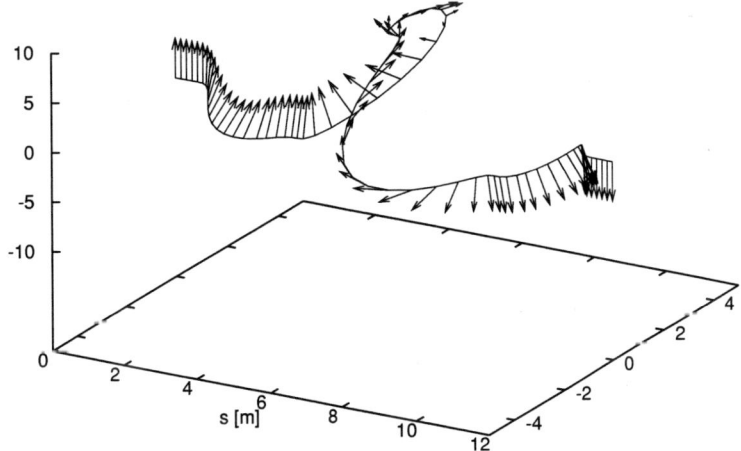

FIGURE 2. Orbit and spin tracking through the four helical magnets of a Siberian Snake at $\gamma = 25$. The spin tracking shows the reversal of the vertical polarization.

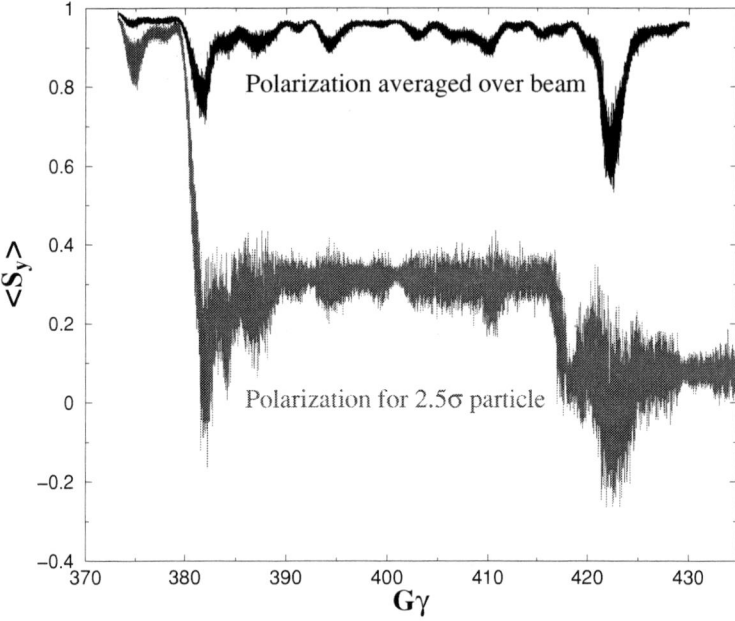

FIGURE 3. Result of spin tracking through the region of the strongest resonance in RHIC. The closed orbit was corrected with residuals of less than 0.2 mm.

ACCELERATING POLARIZED PROTONS IN AGS AND RHIC

Fig. 1 shows the lay-out of the Brookhaven accelerator complex highlighting the components required for polarized beam acceleration. The new 'Optically Pumped Polarized Ion Source' (OPPIS) [4] produced 10^{12} polarized protons per pulse. A single source pulse is captured into a single bunch, which is ample beam intensity to reach the nominal RHIC bunch intensity of 2×10^{11} polarized protons.

In the AGS a 5% solenoidal partial snake that rotates the spin by 9° is sufficient to avoid depolarization from imperfection resonances up to the required RHIC transfer energy of about 25 GeV [5]. Full spin flip at the four strong intrinsic resonances can be achieved with a strong artificial rf spin resonance excited coherently for the whole beam by driving large coherent vertical betatron oscillations [6]. The remaining polarization loss in the AGS is caused by coupling resonances and weak intrinsic resonances. Faster acceleration rate and a future, much stronger partial Snake should eliminate depolarization in the AGS [7].

The full Siberian snakes (two for each ring) and the spin rotators (four provide longitudinal polarization for an experiment) for RHIC each consist of four 2.4 m long, 4 T helical dipole magnet modules each having a full 360° helical twist [8, 9]. The 9 cm diameter bore of the helical magnets can accommodate 3 cm orbit excursions at injection. Fig. 2 shows the orbit and spin trajectory through a RHIC snake. The superconducting helical dipoles were constructed at BNL using thin cable placed into

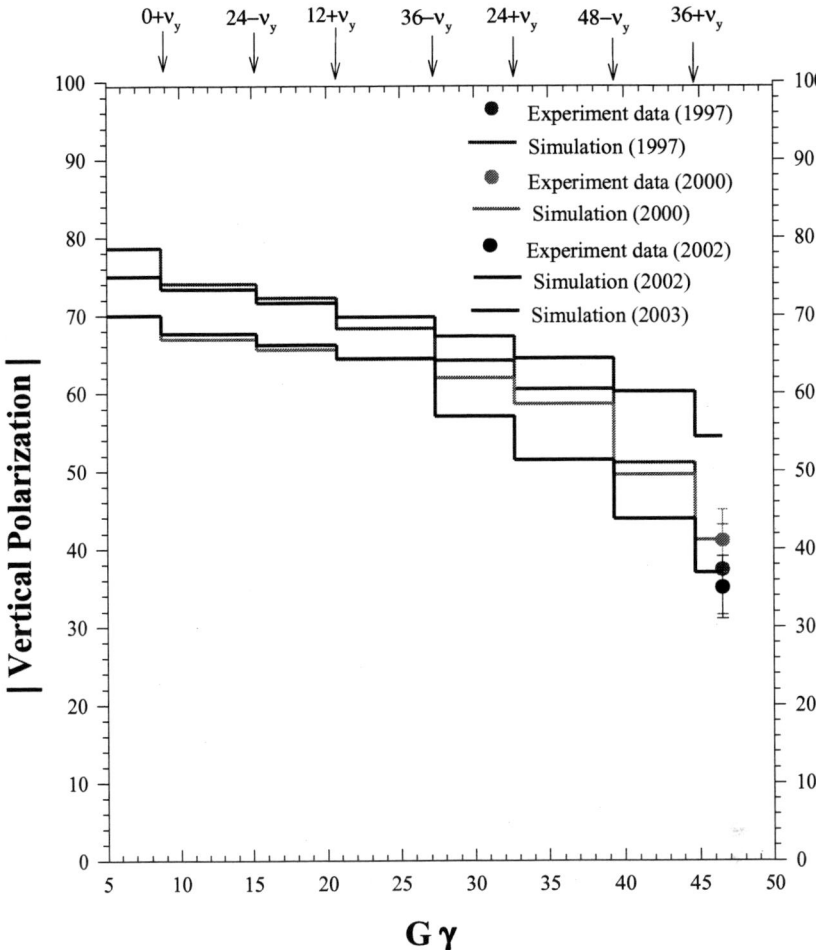

FIGURE 4. Beam polarization measured at the AGS extraction energy for recent running periods. Also shown are simulations of the polarization evolution during the acceleration ramp taking into account all intrinsic depolarizing resonances, the ramp rate, and coupling from the solenoidal partial Siberian snake.

helical grooves that have been milled into a thick-walled aluminum cylinder.

The acceleration through the energy region of the strongest resonance was simulated in great detail including a 1 mm rms misalignment of the quadrupoles, and sextupoles as well as the corrector dipoles used to correct the closed orbit. The result is shown in Fig. 3 for a beam with a normalized 95% emittance of 20π mm mrad. Although there is a significant decrease of the polarization at the energy of the strong resonance, the polarization of the full beam is restored after accelerating completely through the resonance region. The simulation also shows that there is significant polarization loss at

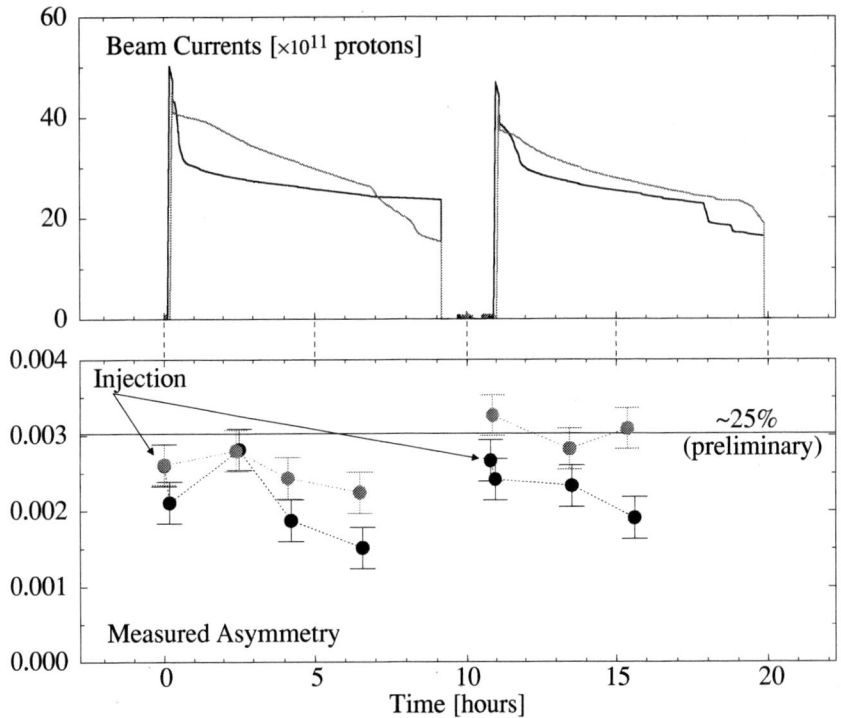

FIGURE 5. Circulating beam and measured asymmetry in the blue and yellow RHIC ring (blue(dark) and yellow(light) lines and symbols, respectively) for two typical stores.

the edge of the beam.

In addition to maintaining polarization, the accurate measurement of the beam polarization is of great importance. Very small angle elastic scattering in the Coulomb-Nuclear interference region offers the possibility for an analyzing reaction with a high figure-of-merit which is not expected to be strongly energy dependent [10]. For polarized beam commissioning in RHIC an ultra-thin carbon ribbon was used as an internal target, and the recoil carbon nuclei were detected to measure both vertical and radial polarization components. The detection of the recoil carbon with silicon detectors using both energy and time-of-flight information showed excellent particle identification. It was demonstrated that this polarimeter can be used to monitor polarization of high energy proton beams in an almost non-destructive manner and that the carbon fiber target could be scanned through the circulating beam to measure polarization profiles.

FIRST RHIC POLARIZED PROTON RUN

The first polarized proton collider run in RHIC took place from Dec. 2001 to Jan. 2002. Polarized beams were successfully accelerated to 100 GeV and stored and collided with

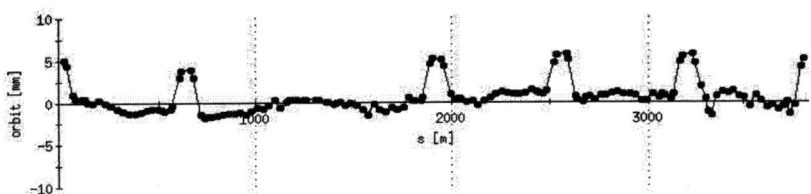

FIGURE 6. Typical vertical closed beam orbit in RHIC during the polarized proton acceleration cycle.

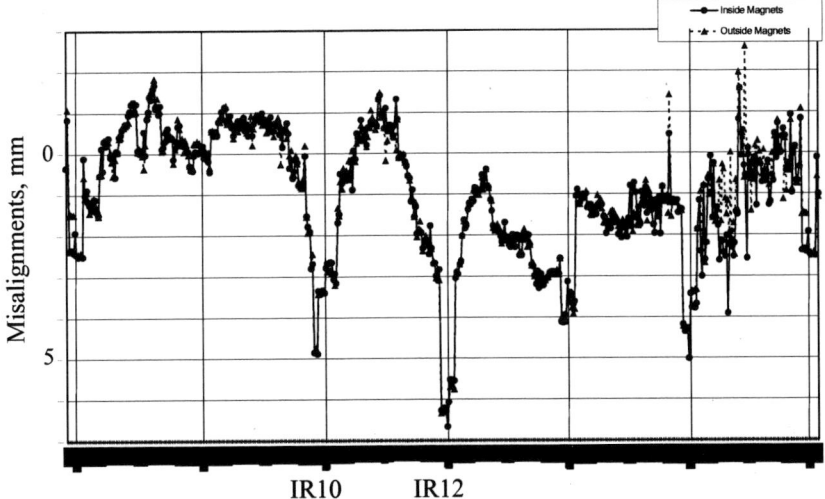

FIGURE 7. Vertical misalignments of the RHIC beam position monitors and quadrupoles

a peak luminosity of about 1.8×10^{30} cm^{-2} s^{-1}.

The beam polarization at the AGS was only about 30% mainly due to the fact that a back-up AGS main power supply had to be used with a much reduced ramp rate that amplified the effect of the weak intrinsic and the coupling depolarizing resonances. Fig. 4 shows the measured beam polarization at AGS extraction energy achieved at recent polarized proton runs. Also shown are simulation results that include the effect of the ramp rate and the coupling from the solenoidal partial Siberian snake on the strength of the intrinsic resonances in the AGS. The simulations are in reasonable agreement with the achieved polarizations. The simulation also predicts more than 50% polarization when the fast ramp-rate AGS main power supply is back in operation again.

In RHIC, essentially all beam polarization was preserved during acceleration and beam storage. Fig. 5 shows circulating beam current and measured asymmetries of two typical stores. The analyzing power at 100 GeV for the RHIC polarimeters is not known but expected to be similar to the value at injection energy. Under this assumption the polarization at store was about 25 %. To preserve beam polarization in RHIC during acceleration and storage the vertical betatron tune had to be maintained between 0.220 and 0.235 and the orbit had to be corrected to better than 1 mm rms. This is in good agreement with the predictions from spin tracking calculations described above.

A typical vertical beam orbit in RHIC is shown in Fig. 6. The beam excursions at the interaction regions serve to avoid beam collisions during acceleration. The remaining part of the orbit was adjusted to minimize vertical orbit excursions in the laboratory frame taking into account the absolute position of the RHIC beam position monitors. A new survey of the vertical alignment of the RHIC beam position monitors was completed after the end of the run and revealed quite large offsets up to about 5 mm as shown in Fig. 7. The effect of this offset on the RHIC beam polarization is still being analyzed, but it is clear that careful vertical realignment of the RHIC magnets and beam position monitors is necessary before acceleration of polarized beam to 250 GeV is attempted. Beam-based orbit flattening could also be successful using vertical dispersion as an indication of vertical orbit errors.

More than 20 years after Y. Derbenev and A. Kodratenko made their proposal to use local spin rotators to stabilize polarized beams in high energy rings, it has now been demonstrated that their concept is working flawlessly even in the presence of strong spin resonances at high energy. Tests, verifying the Siberian Snake concept at low energy, were performed at IUCF [11].

All spin rotators will be installed for the 2003 run with the possibility to go to full collision energy of $\sqrt{s} = 500$ GeV with longitudinal polarization at STAR and PHENIX. Finally a polarized gas jet will be installed for 2004 as an internal target for small angle proton-proton scattering which will allow the absolute calibration of the beam polarization to better than 5 %.

REFERENCES

1. Design Manual - Polarized Proton Collider at RHIC, Brookhaven National Laboratory, July 1998, http://www.rhichome.bnl.gov/RHIC/spin/design .
2. L.H. Thomas, Phil. Mag. **3**, 1 (1927); V. Bargmann, L. Michel, V.L. Telegdi, Phys. Rev. Lett. **2**, 435 (1959).
3. Ya.S. Derbenev et al., Part. Accel. **8**, 115 (1978).
4. A.N. Zelenski et al., "Optically-Pumped Polarized H- ION Sources for RHIC and HERA Colliders", Proc. of PAC99, N. Y., 106 (1999); A.N. Zelenski, these proceedings.
5. H. Huang et al., Phys. Rev. Lett. **73**, 2982 (1994)
6. M. Bai et al., Phys. Rev. Lett. **80**, 4673 (1998); M. Bai, these proceedings.
7. H. Huang, these proceedings.
8. V.I.Ptitsyn and Yu.M.Shatunov, Nucl. Instrum. Methods **A398,** 126 (1997)
9. E. Willen et al., "Construction of helical magnets for RHIC", Proc. of PAC99, N. Y., .3161 (1999).
10. J. Tojo et al., Phys.Rev.Lett. 89, 052302 (2002); O. Jinnouchi et al., SPIN 2002 proceedings (BNL), to be published by AIP; G. Bunce, these proceedings.
11. A. D. Krisch et al., Phys. Rev. Lett. **63**, 1137 (1989); J. E. Goodwin et al., Phys. Rev..Lett. **64**, 2779 (1990)

Thoughts and "Facts" from the AGS Polarized Proton Runs during the 1980's[1]

Leif Ahrens

Brookhaven National Lab

Abstract. This workshop's focus is on considering ways for improving the proton beam polarization that the AGS delivers to the RHIC. This talk attempts to review the first decade of AGS polarization – the 1980's; to briefly describe some aspects of the machine situation, the depolarization avoidance strategies employed and the success achieved in AGS from the perspective of one of those involved.

THE GENERAL CORRECTING SCHEMES OF THE 1980'S

First a very brief description of the 1980's polarized proton setup will be given. Reference [1] goes through this in detail. Differences with the situation in 2003 will be mentioned as we go. The intensity of the polarized beam delivered to the AGS directly from the 200 MeV Linac was at most 2×10^{10} protons per AGS cycle. (In 2003 we will have more than 10 times this intensity, now coming in from the Booster at about 1.5 GeV.) The initial polarization was 75%; we will have 80% this year. The initial transverse emittances of the beam, in AGS, were about 10π mm-mr normalized 95%. The need for "polarization-based" measurements will be indicated occasionally in the following. Machine setup based on polarization measurements are much more time consuming and difficult than either dead reckoned setups or beam-based setups that only need beam properties such as intensity, betatron tune, and transverse emittance, which get more difficult in that order.

Intrinsic Resonances

Intrinsic resonances were handled by pulsing very fast (rise time less than the time for the protons to make one turn around the AGS) ferrite quadrupoles, located symmetrically in each of the twelve AGS superperiods at positions where the vertical betatron function is a maximum (22 m), and the horizontal a minimum (10 m). Only ten quads were actually used, for reasons of cost. Hybrid pulsing systems were built for the resonances occurring at 0+, 12+, 36-, 24+ and 48-, where the code being used here is e.g. "0+" means the resonance when the spin tune ($G\gamma$) is equal to the integer 0 "plus" the vertical betatron tune (which is close to 8.75 in AGS). The resonance at 24- was too weak to require a pulse. The resonance at 36+ ($G\gamma = 44.75$) was judged too

---

[1] Work performed under the auspices of the U.S. Department of Energy

strong for jumping and strong enough to rely on spin flipping. The strengths of these resonances were well predicted by Courant and Ruth. The most aggressive pulsing was at 36-, with the fast (1.6 μsec rise time) components requiring 12 kV, and the slow part (20 μsec) requiring 2 kV. The timing of these pulses during the acceleration ramp was derived from measurements of the field in the main magnets using the AGS "Gauss Clock". Although the timing setups could be "dead reckoned", the clock was neither accurate nor stable enough to avoid using polarization measurements to both carry out the initial setup and to occasionally confirm that things were still ok later in a run. The situation has improved since the 80's. We have a new clock, a new orbit measuring system, and claim (not tested) an accuracy that would allow such a setup to be marginally possible without any polarization checks.

A second system required for the intrinsic tune jumping involved the normal AGS slow quadrupoles. These were used to shift the tunes around each jump to allow more tune headroom for the jump. The timing requirements were mild.

Imperfection Resonances

The imperfection resonances were corrected by making the machine equilibrium orbit perfect for the relevant driving harmonic. There are more than 30 resonances below 18.5 GeV. Each must be set up. The timing requirements are loose enough to be learned from the Gauss clock. Learning the two strength parameters – i.e. the amplitude and the phase of the correction – is completely polarization based. The existing system of vertical correction dipoles, eight per superperiod, was connected to stronger pulsing power supplies using a new control system. The system allowed the harmonic corrections to change with time in order to maximize the current available to the one relevant for the spin survival at that moment in the acceleration cycle. This system was pushed to its limits in order to cope with imperfections encountered below $G\gamma$ of 42. Some of the stronger of these imperfections were ultimately corrected by flipping rather than by correcting since the resulting machine was more stable and the tuning was simple. At highest energies the strength of the correction system was marginal to correct and too weak to flip. Aside from remembering the complexity of the system and the associated setup, this discussion is no longer particularly relevant. The solenoidal Partial Siberian Snake replaced this entire system in the 90's, flipping all the imperfections but also introducing its own interesting problems associated with relatively strong and uncorrectable betatron coupling which then strengthens other higher order resonances.

SPECIFIC DESCRIPTION OF THE MAJOR 1980'S RUNS

The 1980's polarized proton accelerator activity at the AGS can be described by three running periods, a pre-commissioning run in June of 1984, a commissioning run in February of 1986, and a production run in January of 1988. During the 1984 run beam was accelerated up to 16.5 GeV/c or $G\gamma = 31.5$ where 40% polarization was achieved. The higher energy pulsing systems not yet available, the lower energy systems were "being commissioned".

The 1986 Polarized Run

The 1986 run produced the highest energy polarized beam, 45% at 21.7 GeV/c or $G\gamma = 41.5$ of the 80's. (This run is described in reference [1].) The acceleration rate available in the AGS was the nominal 2 Tesla/sec from the Siemens motor generator set (which will not be true for the 1988 run). During the course of the '86 run the correction systems described above were pushed to their limits. Their reliability was better understood, and diagnostic systems to help "Operations" – which for this run was mostly physicists – respond to problems were being developed.

Maintaining the beam intensity and if possible the emittance through the intrinsic jumping was a major effort. The fast quads produce an inherently nonadiabatic change to the orbit betatron motion. Though some concern was expressed over the implications of this for emittance growth due to the implied changes to the machine beta functions, we did not worry about putting the beam on the axis of the quads during the jumps, in particular in the horizontal plane. As a result, each pulsing excited significant oscillations in both the vertical, and the horizontal planes. The shifting of the tunes to provide maximum headroom for the fast tune shift complicated this situation because the horizontal and vertical tunes would sometimes cross slowly – adiabatically – before and after each tune jump. As a result the beam transverse emittances were systematically swapped (slow quad setup), increased (fast quad kicks), and then swapped again (slow quad recovery) around the intrinsic jump. This was not understood at the time though the fascinating transverse size evolution as seen by the ionization profile monitor (IPM) was known. Figure 1 shows the trajectories in tune space of the vertical betatron tune.

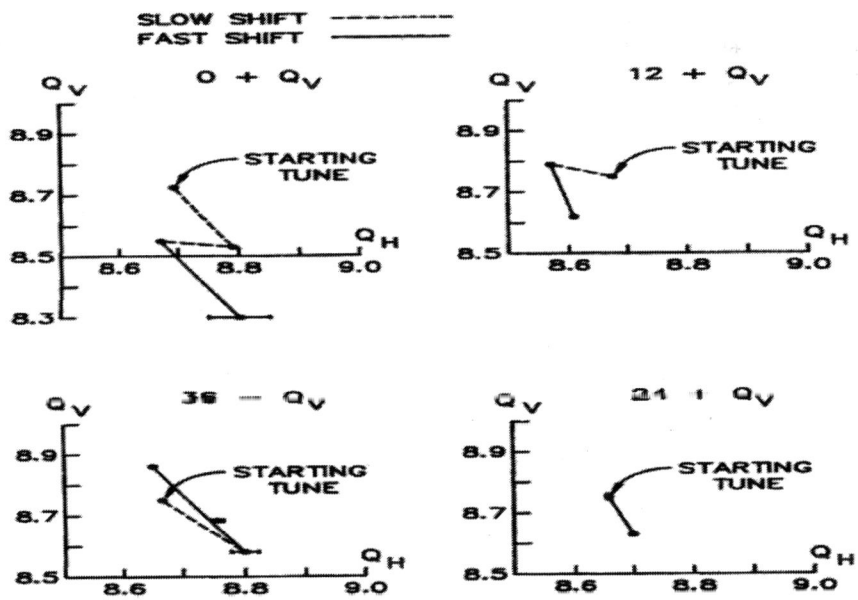

FIGURE 1.   AGS Horizontal Betatron tunes as set up during the 1986 run.

Figure 2 gives one snapshot of the vertical emittance evolution. These figures are from a talk given in 1987 [2]. There is no discussion of the behavior of the horizontal tunes, which must also jump, though only half as far as the vertical and of course are involved in the emittance swap.

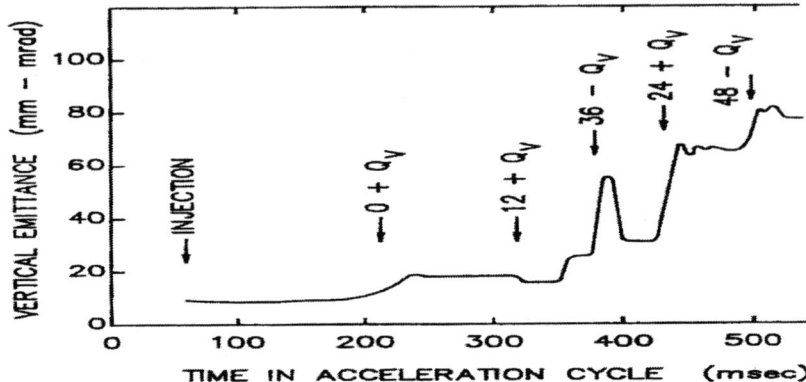

FIGURE 2. Typical vertical normalized emittance (95%) during the 1986 run.

The solution given in these figures was arrived at empirically, to first of all allow beam survival, and then to minimize emittance growth. We well understood that emittance growth was a bad thing for the intrinsics. Note that for the 0+ jump, the vertical tune crossed the half-integer line at 8.5, actually crossed it twice, once very fast going down, and then more slowly on the recovery side of the fast pulse. That this line can be crossed is consistent with earlier AGS experience, at least at injection.

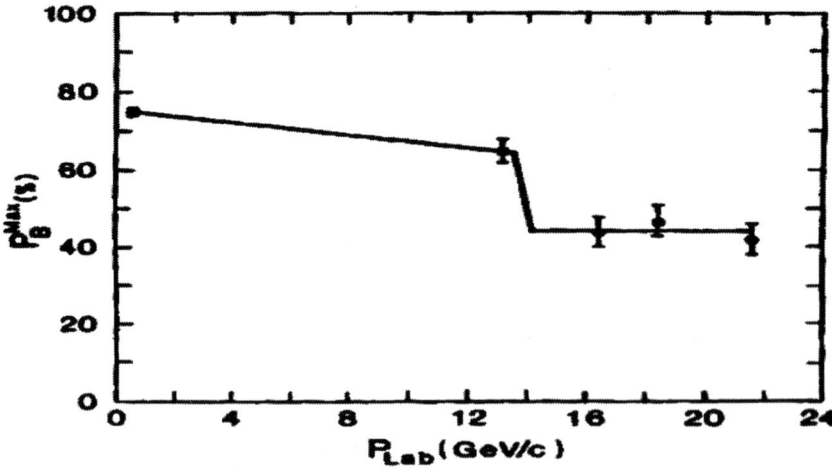

FIGURE 3. Polarization (figure 40 from ref. [1]) measured using extracted beam for the 1986 run.

Figure 3, again from reference [1], gives measurements of the polarization using a polarimeter in an extracted beam line for the 1986 run. The measurement at 13.3 GeV/c corresponds to extraction at Gγ = 25.5, and at 16.5 GeV/c to Gγ = 31.5. The drop between these points is associated with the region near the strong 36- resonance. Aside from this and despite or because of the large emittance growth, there is no measurable polarization loss later in the cycle.

The 1988 Polarized Run

The 1988 run was explicitly to be a production run. We had 2.5 weeks to tune up the machine, and then 3 weeks for physics. Extraction at 18.5 GeV/c (Gγ = 35.5) avoided the higher energy trouble with the imperfection corrections and the 48-intrinsic. Interpretation of the results of the run are complicated by a failure with the Siemens motor-generator set, which forced the run to occur with the lower acceleration rate associated with the backup power supply, the Westinghouse. (If you think I have slipped a decade and am describing the 2002 run, you are wrong, but your confusion is understandable.) So the resonances are all stronger because the acceleration rate is cut in half. In some ways this made the machine setup easier. Tolerances and adjusting room for timing the imperfection resonance corrections and the slow quadrupole tune shifts were relaxed by this factor of two. We had learned enough about the cause for the emittance growth in '86 to fix it. Both the vertical positions (beam based quad repositioning) and the horizontal position (care for the radial position) were corrected. As a result the emittance growth essentially disappeared. Figure 4, from reference [3], compares the emittance growth in the two runs.

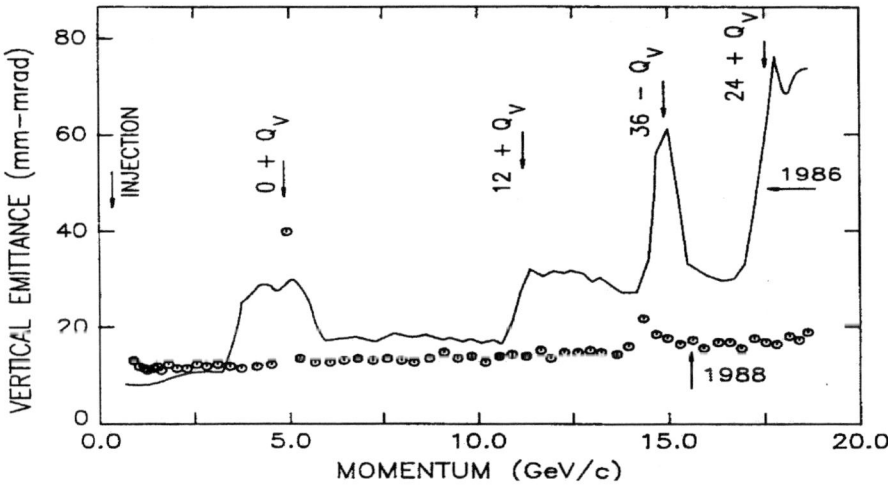

FIGURE 4. Comparison of the vertical emittance growth (normalized, 95%) between 1986 and 1988.

The slow tune shifts to gain jump headroom proved unnecessary except for 36-, and the shift there followed the book, with the vertical tune slowly pushed down to 8.55 and the jump moving it to 8.85.

The systematic tuning done for the 1986 run was repeated for the 1988 run. Despite a resurvey of the ring vertically, the required harmonic corrector strengths to eliminate polarization loss at the imperfections was about the same. Measuring polarization using the internal polarimeter with certainty was a continuing problem. Once we were high enough in momentum to use the external polarimeter, we chose to do so, despite the cost associated with setting up extraction. The measurements were better.

At the end of the tune-up period, beam with polarization of 45% was available for the physics experiment in the extraction line. Over the rest of the run we continued a program of tuning behind the experiment; varying slightly the corrections at the most sensitive resonances while watching the polarization of the extracted beam in an attempt to further increase the delivered polarization. The logbooks, which still exist, display the results of this effort. Scan after scan show the polarization being smoothly optimized above 50%. However, this did not produce any long-term improvement in the beam polarization. At the end of the running period we were still no better than at the beginning. Whether we were missing a critical knob, or were just forever slipping due to tiny changes in the many corrected resonances is not known.

A measurement at the end of the run using the internal polarimeter and collecting data simultaneously over many contiguous gates from just after 0+ till extraction showed only a smoothly falling asymmetry, with no structure to suggest a single point of polarization loss.

ACKNOWLEDGMENTS

These polarization-maintaining systems were complicated to build and to maintain. I would acknowledge the teams of engineers and technicians at BNL, at the University of Michigan, and elsewhere who made them happen.

REFERENCES

1. Khiari, F.Z., et al., "Acceleration of polarized protons to 22 GeV/c and the measurement of spin-spin effects in $p_\uparrow + p_\uparrow \rightarrow p + p$", *Physical Review* **D 39**, 45-85 (1989).
2. Ahrens, L., "Polarized Proton Acceleration at the Brookhaven AGS", in *Proceedings of the 13th International Conference on High Energy Accelerators, Vol 2*, edited by A.N. Skrinsky, Novosibirsk: Publishing House Nauka, Siberian Division (1987), pp. 193-195.
3. Ahrens, L., "Operation of the AGS Polarized Beam" in *Proceedings of the 8th International Symposium on High Energy Spin Physics, Minneapolis 1988*, edited by K. J. Heller, Particle and Fields Series 37, AIP Conference Proceedings **187**, 1068-1076 (1989).

Overcoming Intrinsic and Coupling Spin Resonances in the AGS[1]

M. Bai, L. Ahrens, T. Roser

Brookhaven National Laboratory, Upton, NY 11973, U.S.A

Abstract. In the Brookhaven AGS, polarized protons are accelerated from $G\gamma = 4.5$ to $G\gamma = 46.5$. During the acceleration, a total of 42 imperfection spin depolarization resonances and 7 intrinsic spin resonances are crossed. Currently, the depolarization at each imperfection spin resonance is overcome by a solenoid 5% snake and full spin flips are induced at 4 out of the 7 intrinsic resonances by the AGS rf dipole to avoid the polarization loss. No correction schemes are applied at the remaining 3 weak spin resonances. In addition, coupling spin resonances are also observed due to the solenoidal field of the snake and no correction is applied for these spin resonances other than keeping the horizontal and vertical betatron tunes separated. In order to achieve $\geq 50\%$ beam polarization out of AGS, all of those spin resonances need to be corrected. This paper proposes three correction methods to overcome the strong intrinsic spin resonances as well as the weak intrinsic spin resonances and the coupling spin resonances.

I. INTRODUCTION

In the Brookhaven AGS, the polarized protons are accelerated from $G\gamma = 4.5$ to $G\gamma = 46.5$. A total of 42 imperfection spin resonances [1,2] are crossed during the acceleration. A partial solenoid snake of 5% is used to avoid beam polarization loss at each imperfection resonance [2,3].

Because of the AGS 12-fold symmetry, a total of 7 intrinsic spin resonances are harmful to the beam polarization. The intrinsic spin resonances are $G\gamma = 0 + \nu_z$, $G\gamma = 12 + \nu_z$, $G\gamma = 24 - \nu_z$, $G\gamma = 24 + \nu_z$, $G\gamma = 48 - \nu_z$, $G\gamma = 36 - \nu_z$, and $G\gamma = 36 + \nu_z$ where ν_z is the vertical betatron tune. Among them, the intrinsic resonances at $G\gamma = 0 + \nu_z$, $G\gamma = 12 + \nu_z$ and $G\gamma = 36 \pm \nu_z$ are much stronger than the other three resonances. With the nominal AGS machine and beam conditions, i.e. 10π mm-mrad normalized vertical beam emittance and $\alpha = 4.8 \times 10^{-5}$ acceleration rate, the strong intrinsic resonances can cause total depolarization and even partial spin flips, while the weak spin resonances cause about 5 to 10% polarization losses. Currently, a vertical rf dipole is employed

---

[1] This paper is a summary of M. Bai's three lectures at the Workshop

to induce a full spin flip at each strong intrinsic resonance without losing beam polarization [8]. For the weak spin resonances, no correction is implemented during acceleration.

With the normal AGS polarized proton setup, the horizontal and vertical betatron oscillations are coupled due to the solenoidal field of the 5% snake. Beam depolarization also occurs at the coupling spin resonances $G\gamma = kP \pm \nu_x$ where k is an integer, $P = 12$ is the super-periodicity of the AGS and ν_x is the horizontal betatron tune. The strength of the coupling resonance is proportional to the associated intrinsic resonance as well as the coupling strength. Currently, no correction schemes are applied at these resonances except separating the horizontal and vertical tunes.

In order to achieve more than 50% beam polarization at the AGS extraction energy, one needs to not only fully correct the imperfection spin resonances and the strong intrinsic spin resonances, but also address the beam depolarization at the weak intrinsic resonances and the coupling spin resonances.

II. OVERCOMING STRONG INTRINSIC SPIN RESONANCES WITH AN RF DIPOLE

The intrinsic spin resonance is driven by the quadrupole focusing magnetic fields due to the vertical betatron oscillation. The strength of an intrinsic spin resonance is proportional to the betatron oscillation amplitude. Normally in a beam, particles close to the core of the beam oscillate less than particles around the edge. Thus, the final polarization is an ensemble average of the Froissart-Stora formula over the betatron amplitude of the beam particles. Using the Gaussian beam distribution model, the final polarization becomes

$$P_f = \left(\frac{1 - \pi|\epsilon_{\rm rms}|^2/\alpha}{1 + \pi|\epsilon_{\rm rms}|^2/\alpha}\right) P_i , \qquad (1)$$

where $P_{i,f}$ is the beam polarization before/after crossing the spin resonance. $\epsilon_{\rm rms}$ is the spin resonance strength for a particle with an rms emittance. α is resonance crossing rate given by

$$\alpha = \frac{d(G\gamma - kP \mp m\nu_z)}{d\theta}, \qquad (2)$$

and θ is the orbiting angle in the synchrotron. For a given intrinsic spin resonance, no polarization will be lost if the resonance is crossed very fast, i.e. $\frac{\pi|\epsilon_{\rm rms}|^2}{\alpha} \ll 1$.

In general, there are alternatives to preserve the beam polarization through a spin resonance. One can either "jump" through the resonance so that $\pi|\epsilon_{\rm rms}|^2/\alpha$ is close to zero, or slowly cross the spin resonance so that $\pi|\epsilon_{\rm rms}|^2/\alpha$ is close to infinity to achieve a spin flip. However, for a strong spin resonance, these two techniques are all limited because either the required amount of tune jump and speed of tune jump are not feasible or a very slow ramp rate would cause additional polarized losses from neighboring weak intrinsic spin resonances.

Alternatively, a full spin flip can also be obtained under the normal acceleration rate by enhancing the resonance strength. For intrinsic spin resonances, the effective resonance strength in Eq. (1) can be greatly enhanced in the presence of a large amplitude coherent oscillation, which can be adiabatically excited by an RF dipole. Since the betatron coordinate can be expressed as the linear combination of the vertical betatron motion and the coherent betatron motion [5], the particles experience not only the intrinsic spin resonance, but also a coherent spin resonance at the driving frequency. The resulting polarization, in the limiting case that the driving frequency coincides with the free oscillation frequency, is given by [6]

$$\langle \frac{P_f}{P_i} \rangle = \frac{2}{1+\pi|\epsilon_{\rm rms}|^2/\alpha} \exp\left\{-\frac{(Z_{\rm coh}^2 \hat{\beta}_z/2\beta_z\sigma_z^2)(\pi|\epsilon_{\rm rms}|^2/\alpha)}{1+\pi|\epsilon_{\rm rms}|^2/\alpha}\right\} - 1, \qquad (3)$$

and in the case that the two resonances are well separated, by

$$\langle \frac{P_f}{P_i} \rangle = \frac{1-\pi|\epsilon_{\rm rms}|^2/\alpha}{1+\pi|\epsilon_{\rm rms}|^2/\alpha}\left(2\exp\left\{-\frac{Z_{\rm coh}^2 \hat{\beta}_z}{\beta_z\sigma_z^2}\frac{\pi|\epsilon_{\rm rms}|^2}{2\alpha}\right\} - 1\right). \qquad (4)$$

Here Z_{coh} is the coherent oscillation amplitude, $\hat{\beta}_z$ is the maximum vertical betatron function in the accelerator and σ_z is the rms beam size. Any case in between can produce rich interference patterns and the beam polarization is determined by both the relative strengths and phases of the two resonances [7].

This method has been implemented in the AGS since 1997. The experimental results demonstrated a full spin flip induced by a vertical rf dipole at each strong intrinsic spin resonance [8]. Figure 1 shows the measured polarization at three intrinsic resonances vs the coherent oscillation amplitude which is proportional to the RF dipole strength. The data at spin resonance $12 + \nu_z$ (in the middle plot) demonstrates that the spin was fully flipped at large coherent oscillations where the measured polarization saturated. The same result is also indicated from the data at $0 + \nu_z$ (in the bottom plot) and $36 - \nu_z$ (in the top plot) with the smallest resonance proximity parameter δ. The systematic error of the beam polarization was estimated to be 10%, and the statistical error was about ± 3 %. The lines shown in the figure correspond to results obtained from numerical spin simulations of two spin resonances model.

III. OVERCOMING WEAK INTRINSIC SPIN RESONANCES WITH THE TUNE-JUMP METHOD

For the weak intrinsic spin resonances, in order to achieve a full spin flip using the rf dipole, one needs to excite an extremely large vertical coherence. For the

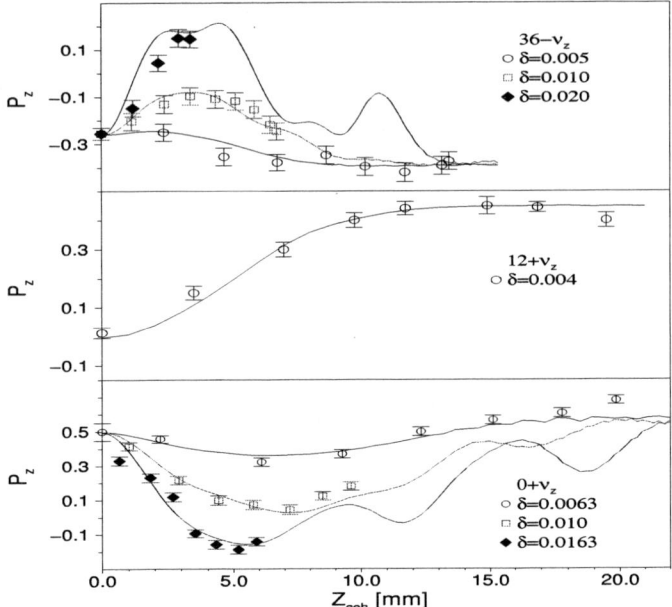

FIGURE 1. The measured proton polarization vs the coherent betatron oscillation amplitude (in mm) for different tune separations at spin depolarizing resonances $0+\nu_z$ (bottom plot), $12+\nu_z$ (middle plot), and $36-\nu_z$ (upper plot) respectively. The error bars show only the statistical errors. The resonance strength of the coherent spin resonance due to the RF dipole is proportional to the coherent betatron amplitude. The lines are the results of multi-particle spin simulations based on the two nearby spin resonances model.

AGS, about $30\sigma_z$ oscillation amplitude is required to induce a full spin flip at $G\gamma = 48 - \nu_z$ where the rms beam size σ_z is 1.4 mm for a beam with vertical emittance of 10π mm-mrad. This certainly exceeds the limit of the AGS physical aperture $(6''(H) \times 3''(V))$. However compared with strong intrinsic resonances, the much weaker strength allows one to be able to preserve the beam polarization with a moderate tune jump through the weak intrinsic resonances.

The principle of the tune-jump technique is to speed up the resonance crossing rate by pulsing a family of quadrupoles within a very short time. The schematic drawing in Fig. 2 shows how the tune-jump method works. The effective resonance crossing rate α_{tj} become

$$\alpha_{tj} = \alpha + \frac{|\Delta\nu_z|}{\Delta\theta}, \qquad (5)$$

where α is the original resonance crossing rate due to acceleration, $\Delta\nu_z$ is the total amount of tune-jump and $\Delta\theta$ is the tune-jump duration. The polarization of a

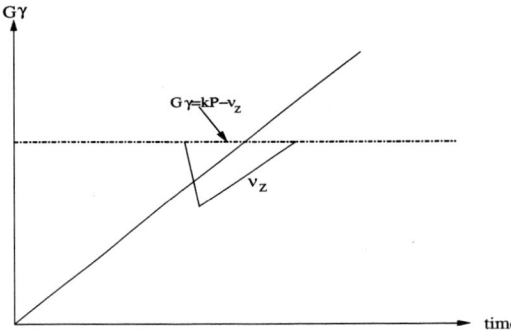

FIGURE 2. Schematic plot of tune-jump through a intrinsic resonance at $G\gamma = kP - \nu_z$. For the intrinsic resonance at $G\gamma = kP + \nu_z$, the tune jump will have the opposite polarity.

Gaussian distributed beam after tune-jumping through an intrinsic spin resonance is given by Eq. 1.

In order to preserve the beam polarization through a weak intrinsic resonance, the size of the tune jump $\Delta\nu_z$ and the duration of the tune jump $\Delta\theta$ are the two critical parameters. Normally, the size of the tune jump is proportional to the strength of the resonance and the duration of the jump is determined by the desired resonance crossing rate based on Eq. 1. Table 1 lists the three weak spin resonances' strength for a beam with vertical 10π mm-mrad normalized emittance as well as the tune-jump size and duration. Here, the tune-jump speed for the weak

TABLE 1. AGS weak intrinsic spin resonance strengths

P[GeV/c]	$G\gamma$	ϵ_k	$\Delta\nu_z$	jump time	$\frac{p_f}{p_i}$
7.95	$24 - \nu_z$	0.0002	0.11	1.89 ms	0.99
17.08	$24 + \nu_z$	0.0004	0.11	473 μs	0.99
20.54	$48 - \nu_z$	0.0006	0.11	210 μs	0.99

intrinsic resonance at $G\gamma = 24 - \nu_z$ is less than the AGS nominal acceleration rate with Siemens power bank and no correction needs to be done for this resonance during the AGS normal operation.

This technique has been successfully demonstrated during the 1986 AGS polarized proton acceleration experiment [9]. A total of ten ferrite quadrupoles were pulsed at all the intrinsic spin resonances between the AGS injection energy of 200 MeV and extraction energy of 22 GeV, except for $G\gamma = 24 - \nu_z$. Each pulsed quadrupole is 50 cm long and is capable of reaching a maximum field gradient of 11.7 kG/m. The fast quadrupoles were designed to have a rise time of 1.6μs. In order to achieve such a fast time constant, ceramic beam pipes without any coating were used in these magnets. The elliptical aperture of the ceramic pipe is 8.5 cm(H)x7.4 cm(V). During the experiment, the maximum achieved tune jump was 0.28 in 1.6μs.

Although this method was used to overcome the strong resonances at $G\gamma = 0+\nu_z$, $G\gamma = 12 + \nu_z$ and $G\gamma = 36 - \nu_z$ during the experiment, the large tune-jump size (0.28) and incredible speed at strong spin resonance not only brought a great difficulty in system implementation but also caused beam emittance growth. The ceramic beam pipes later on became a problem for the AGS high intensity proton operation because of their high impedance and vertical aperture restrictions, and all the fast quadrupoles were later removed from the beam line.

Since the required tune-jump size and speed is significantly less for the weak spin resonances, this method is more suitable for crossing weak spin resonances. The moderate tune-jump size allows one to be able to afford a fast quadrupole with larger aperture. The much slower crossing speed also allows one to put in aluminum strips through the ceramic beam pipe to reduce the impedance for the AGS high intensity operation. For the AGS, $G\gamma = 48 - \nu_z$ is the strongest among the three weak intrinsic spin resonances. To achieve the tune-jump size listed in the Table 1, the desired quadrupole gradient is 5.5 kG/m assuming 10 fast quadrupoles distributed at the maximum vertical beta function (25 m) around the ring. Although to obtain a tune-jump time of 210 μs, a ceramic beam pipe is inevitable, a set of aluminum strips can be placed outside the beam pipe to provide a path for the image current and thus reduce the impedance.

IV. COUPLING SPIN RESONANCES

In a perfect accelerator, the horizontal and vertical betatron oscillations are independent of each other. However, this independence can be broken if there is any quadrupole roll error or solenoid field. In this case, the horizontal motion then gets coupled to the vertical oscillation. Unlike the uncoupled case, the frequency spectrum of the betatron oscillation in either of the two transverse plane then consists of two components ν_1 and ν_2 given by [4,10]

$$\nu_1 = \frac{1}{2}(\nu_x + \nu_z) + \frac{1}{2}\sqrt{(\nu_x - \nu_z)^2 + \Delta Q_{min}^2} \to \nu_x; \text{ without coupling} \quad (6)$$

$$\nu_2 = \frac{1}{2}(\nu_x + \nu_z) - \frac{1}{2}\sqrt{(\nu_x - \nu_z)^2 + \Delta Q_{min}^2} \to \nu_z; \text{ without coupling} \quad (7)$$

where ν_x and ν_z are the unperturbed horizontal and vertical tune. ΔQ_{min} is the minimum tune split between the two eigen tunes when $\nu_x = \nu_z$. It is given by

$$\Delta Q_{min} = \frac{1}{2\pi}\oint \sqrt{\beta_x\beta_z} A_{xz} e^{i(\nu_x\phi_x - \nu_z\phi_z - (\nu_x-\nu_z-l)\frac{s}{R})}ds, \quad (8)$$

where $\beta_{x,z}$ are the betatron amplitude functions for the horizontal and vertical planes and A_{xz} is proportional to the strength of the coupling elements [11]. For a solenoid magnet,

$$A_{xz} = \frac{B_{//}}{2B\rho}[(\frac{\alpha_x}{\beta_x} - \frac{\alpha_z}{\beta_z}) + i(\frac{1}{\beta_x} + \frac{1}{\beta_z})], \qquad (9)$$

where $B_{//}$ is the solenoid magnetic field strength and $B\rho$ is the momentum per charge. With weak coupling,

$$\nu_1 \simeq \nu_x \qquad (10)$$

$$\nu_2 \simeq \nu_z. \qquad (11)$$

In the Brookhaven AGS, the main coupling source comes from the solenoid partial snake which is used to overcome the imperfection spin resonances in the AGS [2]. The snake is 2.4384 m long and located in the $I10$ straight section where $\beta_x = 12.2$ m, $\beta_z = 18.2$ m, $\alpha_x = 1.050$ and $\alpha_z = -1.470$. For a 5% snake, $\frac{B_{//}}{B\rho} = 0.023$ m^{-1}, and the minimum tune split ΔQ_{min} from a 5% partial snake is

$$\Delta Q_{min} = 0.01435. \qquad (12)$$

In a coupled machine, in addition to the intrinsic spin resonance at $G\gamma = kP \pm \nu_2$ ($G\gamma = kP \pm \nu_z$ without coupling) [1], the vertical betatron oscillation also drives a coupling spin resonances at $G\gamma = kP \pm \nu_1$. The strength of the coupling resonance ϵ_{ν_x} is proportional to the amount of the coupling and is given by

$$\epsilon_{\nu_x} \propto C_x \sqrt{\varepsilon_u} \epsilon_{\nu_z} \qquad (13)$$

where ϵ_{ν_z} is the strength of the adjacent intrinsic spin resonance and C_x is the coupling coefficient. For a fully coupled machine, $\nu_x = \nu_z$ and $C_x = 1$. For a decoupled machine, $C_x = 0$. ε_u is the beam emittance in the eigen direction [12] and equals the horizontal beam emittance if $C_x = 0$.

Traditionally in the AGS, the beam polarization loss at the coupling resonances is minimized by separating the horizontal and vertical set points. The coupling resonances around these four strong intrinsic spin resonances can produce about 35% polarization losses with the normal AGS polarized proton setting [3,13]. In order to achieve 70% polarization in the AGS, one needs to minimize the polarization loss at the coupling resonances. Because they are adjacent to the intrinsic resonances, it is very difficult to use the vertical rf dipole [8] to obtain two full spin flips at both the intrinsic and the coupling resonances.

IV.A Using a Horizontal Rf Dipole to Cross the Intrinsic and Coupling Resonances

Analogous to the method of using a vertical rf dipole at the intrinsic spin resonance, one should also expect to obtain a full spin flip by inducing a strong artificial resonance if the intrinsic and its coupling spin resonances are fully overlapped. Because of the coupling effect, the two spin resonance can never be brought closer than the minimum tune split ΔQ_{min}. However, ΔQ_{min} in general is small and a full spin-flip still should be achievable if the induced resonance is strong enough. In a fully coupled machine, the unperturbed tunes are equal and the intrinsic and the coupling resonances are equally strong and located on either side of the unperturbed betatron tune at a distance of half of ΔQ_{min}.

Unlike using a vertical rf dipole to obtain a vertical coherence in an uncoupled machine [5], the vertical coherence is excited by a horizontal rf dipole in a fully coupled machine. To understand this, let us first transform the beam motion from the normal geometric coordinates (x, z, s) to the (u, v, s) coordinate system in which the betatron motions along the two eigen directions are fully decoupled [12], i.e.

$$\begin{pmatrix} x \\ x' \\ z \\ z' \end{pmatrix} = R \begin{pmatrix} u \\ u' \\ v \\ v' \end{pmatrix}. \qquad (14)$$

where R is the transformation matrix between (x, z, s) and (u, v, s). In a fully coupled machine with weak coupling coming from a solenoid magnet, the minimum tune split $\Delta Q_{min} \ll 1$ and the R matrix is given by (See the Appendix)

$$R = \frac{1}{\sqrt{2}} \begin{pmatrix} 1 & 0 & -a(\alpha_x + \alpha_z) & -a(\beta_x + \beta_z) \\ 0 & 1 & a(\gamma_x + \gamma_z) & a(\alpha_x + \alpha_z) \\ -a(\alpha_x + \alpha_z) & -a(\beta_x + \beta_z) & 1 & 0 \\ a(\gamma_x + \gamma_z) & a(\alpha_x + \alpha_z) & 0 & 1 \end{pmatrix}, \qquad (15)$$

where $a = \frac{1}{\sqrt{\beta_x \gamma_z + \beta_z \gamma_x + 2(1-\alpha_x\alpha_z)}}$ and $\beta_{x,z}$ are the horizontal and vertical betatron functions at the position of the solenoid magnet. $\gamma_{x,z} = \frac{1+\alpha_{x,z}^2}{\beta_{x,z}}$ and $\alpha_{x,z} = -\frac{1}{2}\beta'_{x,z}$ are the corresponding twiss parameters. Here $'$ is the derivative with respect to the longitudinal coordinate s. With the weak coupling force, one also has [14]

$$\beta_u \simeq \beta_x; \quad \beta_v \simeq \beta_z \qquad (16)$$
$$\alpha_u \simeq \alpha_x; \quad \alpha_v \simeq \alpha_z \qquad (17)$$

where $\beta_{u,v}$ and $\alpha_{u,v}$ are the Courant-Snyder parameters in the two eigen directions.

With a horizontal rf dipole $\Delta B_y L = \Delta B_{ym} L \cos\nu_m\theta$, the horizontal excitation is $\delta x' \cos\nu_m\theta = \frac{\Delta B_{ym}L}{B\rho}\cos\nu_m\theta$, where $B\rho$ is the magnetic rigidity. The corresponding excitation in the (u, v) coordinates are

$$\delta u = 0 \qquad (18)$$
$$\delta u' = \frac{1}{\sqrt{2}}\delta x' \qquad (19)$$
$$\delta v = \frac{a}{\sqrt{2}}(\beta_x + \beta_z)\delta x' \qquad (20)$$
$$\delta v' = -\frac{a}{\sqrt{2}}(\alpha_x + \alpha_z)\delta x'. \qquad (21)$$

Similar to the case of using a vertical rf dipole in an uncoupled machine [5], the coherent betatron motion is a fixed point in the frame which rotates with the rf dipole modulation frequency as shown in Fig. 3. During every modulation period,

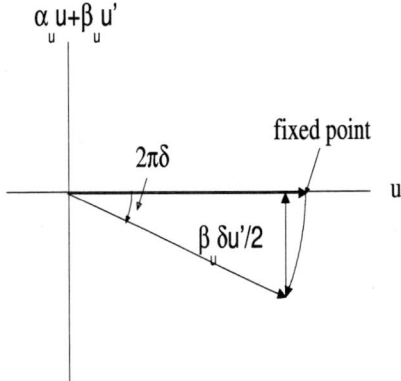

FIGURE 3. Derivation of the coherent betatron motion in the (u, u') plane in the rotating frame. In each modulation period, the phase-space vector rotates through an angle of $2\pi\delta$ and is then given an effective kick of $\frac{\beta_u \delta u'}{2}$ by the rf dipole field.

the particle's phase-space vector rotates through an angle of $2\pi\delta$ in the rotating frame where $\delta = |\nu_m - \nu_{1,2}|$. It then gets deflected by the rf dipole. The effective deflection is $\frac{1}{2}\beta_u \delta u'$ due to the fact that the deflection given by the rf dipole is oscillating at the same frequency. Hence, the fixed point in the (u, u') plane is at

$$f_u = b \frac{a}{\sqrt{2}} \beta_x \delta x' \tag{22}$$

$$f_{u'} = -\frac{\alpha_x}{\beta_x} b \frac{a}{\sqrt{2}} \beta_x \delta x' \tag{23}$$

where $b = \frac{1}{4\pi\delta}$. The fixed point in the (v, v') plane can be calculated using the same method. Since $\delta v \neq 0$, the deflection given by the rf dipole in the (v, v') plane is not along the axis of $\alpha_v v + \beta_v v'$. The fixed point in the (v, v') plane is given by

$$f_v = -b \frac{a}{\sqrt{2}} (\alpha_z \beta_x - \beta_z \alpha_x) \delta x' \tag{24}$$

$$f_{v'} = b \frac{a}{\sqrt{2}} (\beta_x \gamma_z - \alpha_x \alpha_z + 1). \tag{25}$$

Transforming the fixed points in the (u, v) coordinate system back to the normal geometric (x, z) coordinates,

$$f_x/b = \frac{1}{\sqrt{2}} f_u - \frac{a}{\sqrt{2}} (\alpha_x + \alpha_z) f_v - \frac{a}{\sqrt{2}} (\beta_x + \beta_z) f_{v'}$$

$$f_{x'}/b = \frac{1}{\sqrt{2}} f_{u'} + \frac{a}{\sqrt{2}}(\gamma_x + \gamma_z) f_v + \frac{a}{\sqrt{2}}(\alpha_x + \alpha_z) f_{v'}$$
$$f_z/b = -\frac{a}{\sqrt{2}}(\alpha_x + \alpha_z) f_u - \frac{a}{\sqrt{2}}(\beta_x + \beta_z) f_{u'} + \frac{1}{\sqrt{2}} f_v$$
$$f_{z'}/b = \frac{a}{\sqrt{2}}(\gamma_x + \gamma_z) f_u + \frac{a}{\sqrt{2}}(\alpha_x + \alpha_z) f_{u'} + \frac{1}{\sqrt{2}} f_{v'} \qquad (26)$$

By substituting Eq. 25 to Eq. 26, one then obtains

$$f_x/b = 0$$
$$f_{x'}/b = 0$$
$$f_z/b = a(\beta_z \alpha_x - \beta_x \alpha_z)\delta x'$$
$$f_{z'}/b = a(\beta_x \gamma z - \alpha_x \alpha_z + 1)\delta x'. \qquad (27)$$

This demonstrates that in a fully coupled machine, a vertical coherence can be excited by applying a horizontal rf dipole. The amplitude of the vertical coherence is $\frac{BL}{4\pi B\rho\delta}\sqrt{\beta_x\beta_z}$.

Fig. 4 shows the numerical spin tracking results at $G\gamma = 36 + \nu_z$. The dotted line shows the result with the nominal AGS tune setting ($\nu_x = 8.8, \nu_z = 8.7$) and no correction scheme for the intrinsic spin resonance. In this case, the depolarization at the coupling resonance is obvious. The solid line is the result of using a horizontal rf dipole with the horizontal and vertical betatron tunes set at 8.7. Due to the coupling from the solenoid partial snake, the two betatron tunes are split by 0.0144. The horizontal rf dipole tune was set at 0.3. With a horizontal RF dipole amplitude 28.0 G-m, a full spin flip was achieved.

IV.B Experimental Results

The method of using a horizontal rf dipole to excite a vertical coherence to cross the coupling spin resonance was tested in the AGS during the 2000 RHIC polarized proton commissioning run. The polarized H^- beam was pre-accelerated in the 200 MeV LINAC and then stripped and injected into the Booster. It was then injected into the AGS at $G\gamma = 4.7$ and accelerated up to $G\gamma = 46.5$. In the AGS, the nominal tune setting is $\nu_x = 8.8$ and $\nu_z = 8.7$.

During the experiment, the AGS skew quadrupoles were all set to 17 A. Due to a hardware limit, the partial snake strength at $G\gamma = 36 + \nu_z$ was actually only about 3.5% instead of 5%. The combined effect of the skew quadrupoles and the weaker snake gave a smaller minimum tune spilt ΔQ_{min} of 0.007. The horizontal rf dipole was set in the middle of the two betatron tunes ν_1 and ν_2. The turn-by-turn beam position monitor data confirmed that a vertical coherence was excited without horizontal response as shown in the left part of Fig. 5. The horizontal response was not zero once the rf dipole tune deviated from the average of the two eigen tunes as shown on the right of Fig. 5.

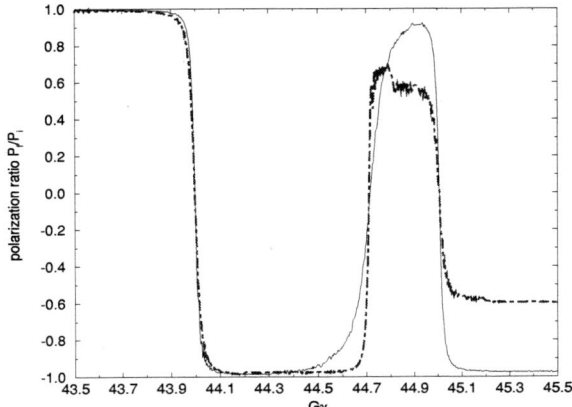

FIGURE 4. This figure shows the calculated polarization ratio P_f/P_i as a function of energy. The solid line is for the case of a fully coupled machine and a horizontal rf dipole used to obtain an adiabatic vertical coherence. The dotted line is the result of a weakly coupled machine with the two betatron tunes set 0.1 apart. No correction scheme was used at the $G\gamma = 36 + \nu_z$ intrinsic spin resonance. For both cases, the horizontal and vertical emittance are 20π mm-mrad and 10π mm-mrad respectively.

Table 2 shows the comparison of the measured beam asymmetries of using vertical rf dipole, no correction and using horizontal rf dipole at $G\gamma = 36 + \nu_z$. Comparing the measured asymmetry when using the horizontal rf dipole with the case of no correction, it is clear that the horizontal rf dipole did help to recover the beam polarization. However, the excited coherence was not optimized and about 70%

TABLE 2. Measured beam asymmetries

	Measured asymmetry ($\times 10^{-3}$)	Condition
1	1.50± 0.04	with vertical rf dipole
2	1.25± 0.1	with horizontal rf dipole
3	0.067± 0.063	no correction

beam emittance growth was observed. Because of limitations of the AGS sextupole power supplies, we could not achieve small chromaticities in both planes and obtain a fully adiabatic excitation. This was the most likely reason that the horizontal rf dipole did not recover 100% beam polarization as expected.

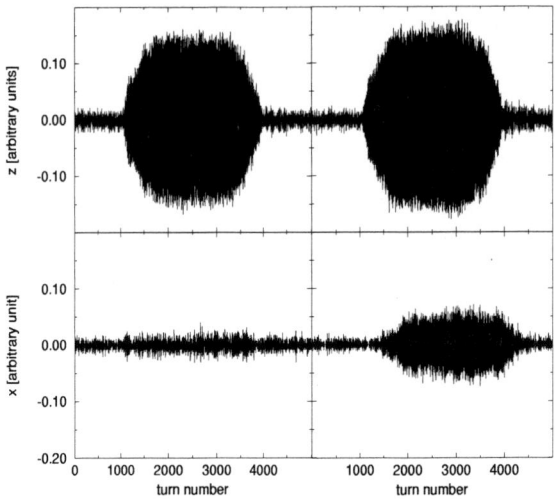

FIGURE 5. The top and bottom plots on the left are vertical and horizontal turn-by-turn beam position data when the horizontal rf dipole modulation tune $\nu_m = \frac{1}{2}(\nu_1 + \nu_2)$. As shown, no horizontal coherence was excited. The two plots on the right correspond to the case where the horizontal modulation tune $\nu_m \neq \frac{1}{2}(\nu_1 + \nu_2)$ and the horizontal coherence was no longer zero.

V. ACKNOWLEDGMENTS

The three techniques discussed in this report were done in the AGS over several years and several papers have been published in various journals as well as conference proceedings. The authors would like to thank all the people who have been involved in these research. In addition, we would also like to thank Dr. E. D. Courant, W. J. Glenn, Dr. H. Huang, Dr. A. Lehrach, Dr. A. Luccio, Dr. W. Mackay, V. Ranjbar, Dr. N. Tsoupas, Dr. W. van Asselt for the fruitful discussions. We also would like to thank K. Zeno and D. Warburton for their great help. This work is performed under the auspices of Department of Energy of U.S.A.

VI. APPENDIX

In an accelerator with a solenoid magnet as the only source of coupling, the one-turn matrix T is

$$T = \begin{pmatrix} M & n \\ m & N \end{pmatrix} = \begin{pmatrix} p & q \\ -q & p \end{pmatrix} \begin{pmatrix} M_1 & 0 \\ 0 & N_1 \end{pmatrix}. \qquad (28)$$

where M_1 and N_1 are the matrices starting from the beginning of the solenoid to the end of the solenoid. M and N are the horizontal and vertical matrices.

$$M = \begin{pmatrix} cos\mu_x + \alpha_x sin\mu_x & \beta_x sin\mu_x \\ -\gamma_x sin\mu_x & cos\mu_x - \alpha_x sin\mu_x \end{pmatrix} \tag{29}$$

and

$$N = \begin{pmatrix} cos\mu_z + \alpha_z sin\mu_z & \beta_z sin\mu_z \\ -\gamma_z sin\mu_z & cos\mu_z - \alpha_z sin\mu_z \end{pmatrix} \tag{30}$$

where $\alpha_{x,z}$, $\beta_{x,z}$ and $\gamma_{x,z}$ are the horizontal and vertical Courant-Snyder parameters. $\mu_{x,z}$ are the horizontal and vertical phase advances. p and q are the components in the solenoid transfer matrix T_s.

$$T_s = \begin{pmatrix} p & q \\ -q & p \end{pmatrix} = \begin{pmatrix} C^2 & \frac{SC}{K} & SC & \frac{S^2}{K} \\ -KSC & C^2 & -kS^2 & SC \\ -SC & -\frac{S^2}{K} & C^2 & \frac{SC}{K} \\ kS^2 & -SC & -KSC & C^2 \end{pmatrix}, \tag{31}$$

where $K = \frac{B_{//}}{2B\rho}$, $C = cos(KL)$, $S = sin(KL)$ and L is the length of the solenoid. $B_{//}$ is the strength of the solenoid. From Eq. 28, one has

$$pM_1 = M; \tag{32}$$
$$pN_1 = N; \tag{33}$$
$$n = qN_1 = qp^{-1}N; \tag{34}$$
$$m = -qM_1 = qp^{-1}M; \tag{35}$$
$$\tag{36}$$

In a fully coupled machine where $\mu_x = \mu_z$, the transformation matrix R which diagonalizes the one turn matrix T in Eq. 28 is given by [2]

$$R = \begin{pmatrix} \sqrt{1 - |E|} & -sE\ s \\ -E & \sqrt{1 - |E|} \end{pmatrix} \tag{37}$$

where

$$s = \begin{pmatrix} 0 & 1 & 0 & 0 \\ -1 & 0 & 0 & 0 \\ 0 & 0 & 0 & 1 \\ 0 & 0 & -1 & 0 \end{pmatrix} \tag{38}$$

and

$$E = -\frac{\sqrt{2}}{4sin\mu sin(\pi\Delta Q_{min})}(m - sn\ s). \tag{39}$$

where ΔQ_{min} is the minimum tune split given by [11]

$$\Delta Q_{min} = \frac{KL}{2\pi}\sqrt{\beta_x\beta_z}[(\frac{\alpha_x}{\beta_x} - \frac{\alpha_z}{beta_z}) + i(\frac{1}{\beta_x} + \frac{1}{\beta_z})], \qquad (40)$$

where $KL = \frac{B_{//}L}{2B\rho}$. The size of the minimum tune split is

$$\Delta Q_{min} = \frac{KL}{2\pi}a = \frac{KL}{2\pi}\sqrt{\beta_x\gamma_z + \beta_z\gamma_x + 2(1 - \alpha_x\alpha_z)}. \qquad (41)$$

In a fully coupled machine with weak coupling strength, namely $\Delta Q_{min} \ll 1$ and $KL \ll 1$, we then have

$$sin(\pi\Delta Q_{min}) \simeq \pi\Delta Q_{min}, \qquad (42)$$

and

$$\frac{sin(KL)}{cos(KL)} \simeq KL. \qquad (43)$$

Substituting Eq. 42, Eq. 43, m in Eq. 36 and n in Eq. 35 into Eq. 39, one then gets

$$E = -\frac{1}{\sqrt{2}}a\begin{pmatrix} -(\alpha_x + \alpha_z) & -(\beta_x + \beta_z) \\ (\gamma_x + \gamma_z) & (\alpha_x + \alpha_z) \end{pmatrix}. \qquad (44)$$

Hence, the R matrix is

$$R = \frac{1}{\sqrt{2}}\begin{pmatrix} 1 & 0 & -a(\alpha_x + \alpha_z) & -(\beta_x + \beta_z) \\ 0 & 1 & (\gamma_x + \gamma_z) & (\alpha_x + \alpha_z) \\ -a(\alpha_x + \alpha_z) & -(\beta_x + \beta_z) & 1 & 0 \\ (\gamma_x + \gamma_z) & (\alpha_x + \alpha_z) & 0 & 1 \end{pmatrix} \qquad (45)$$

and its inverse matrix R^{-1} is

$$R^{-1} = \frac{1}{\sqrt{2}}\begin{pmatrix} 1 & 0 & a(\alpha_x + \alpha_z) & (\beta_x + \beta_z) \\ 0 & 1 & -(\gamma_x + \gamma_z) & -(\alpha_x + \alpha_z) \\ a(\alpha_x + \alpha_z) & (\beta_x + \beta_z) & 1 & 0 \\ -(\gamma_x + \gamma_z) & -(\alpha_x + \alpha_z) & 0 & 1 \end{pmatrix} \qquad (46)$$

VII. REFERENCES

1. E. D. Courant, R. D. Ruth, *The Acceleration of Polarized Protons in Circular Accelerators*, BNL report, BNL 51270, 1980.
2. T. Roser, *Partial Siberian Snake Test at the Brookhaven AGS*, in High Energy Spin Physics: 10th International Symposium, ed. T.Hasegawa, et al., Nagoya, Japan, 1992 (Universal Academic Press, Inc.,1992), p. 429.
3. H. Huang et al., *Preservation of Proton Polarization by a Partial Siberian Snake*, Phys. Rev. Letters **73**, 2982 (1994).

4. S. Y. Lee, *Accelerator Physics*, World Scientific Pub. Singapore, 1999.
5. M. Bai, et al., *Experimental Test of Coherent Betatron Resonance Excitations*, Physical Review E **5**, 6002 (1997).
6. M. Bai, S. Y. Lee, H. Huang, T. Roser and M. Syphers, AGS/RHIC/SN No. 055.
7. S. Tepikian, S. Y. Lee, E. D. Courant, Particle Accelerators **20**, 1 (1986).
8. M. Bai et al., *Overcoming Intrinsic Spin Resonances with an rf Dipole*, Physical Review Letters **80**, 4673(1998).
9. F. Z. Khiari et al., *Acceleration of polarized protons to 22 GeV/c and the measurement of spin-spin effects in $p_\uparrow + p_\uparrow \to p + p$*, Physical Review D, **39**, p. 45(1989).
10. D. A. Edwards, M. J. Syphers, *An Introduction To The Physics of High Every Accelerators*, Wiley-Interscience Pub. 1993.
11. S. Y. Lee, *Spin Dynamics and Snakes in Synchrotrons*, World Scientific Pub. Singapore, 1997.
12. D. A. Edwards, L. C. Teng, *Parametrization of LINEAR Coupled Motion in Periodic Systems*, IEEE Trans. on Nucl. Sc. **20**, 885 (1973).
13. H. Huang et al., *Polarized Proton Beam in the AGS*, Proceedings of 13th international symposium in High Energy Spin Physics, p. 492 (1998).
14. T. Roser, *Multiturn Injection With Coupling*, AGS/AD/Tech. Note No. 354.

Overcoming Depolarizing Resonances at COSY

A. Lehrach, U. Bechstedt, J. Dietrich, R. Gebel, B. Lorentz, R. Maier,
D. Prasuhn, A. Schnase, H. Schneider, R. Stassen, H. Stockhorst, R. Tölle

Forschungszentrum Jülich GmbH, 52425 Jülich, Germany

Abstract. At the cooler synchrotron COSY, polarized protons and deuterons are accelerated up to 3.65 GeV/c. In a strong-focusing synchrotron like COSY, imperfection and intrinsic resonances cause polarization losses during acceleration. During the acceleration of polarized protons, five imperfection resonances are crossed. The existing magnet system of COSY allows to overcome all imperfection resonances by exciting adiabatic spin flips without polarization losses. The number of intrinsic resonances depends on the superperiodicity of the lattice. A tune-jump system consisting of two fast quadrupoles has been developed to handle intrinsic resonances in COSY. Vertically polarized proton beams with more than 75% polarization have been delivered in recent years to internal as well as to external experiment areas at different momenta up to the maximum momentum of COSY. For the acceleration of polarized deuterons, additional correction provisions are not necessary to preserve polarization during acceleration.

In this paper the techniques to overcome depolarizing resonances and the status of the polarized beam acceleration at COSY are discussed.

INTRODUCTION

The COoler SYnchrotron and storage ring COSY at the Forschungszentrum Jülich accelerates polarized protons and deuterons to momenta between 600 MeV/c and 3.65 GeV/c [1, 2]. A polarized ion source originally developed by a collaboration of the universities of Bonn, Erlangen, and Cologne [3] provides vector polarized proton beam and all possible combinations of vector and tensor polarized deuteron beams [4]. The polarized H^- or D^- ion beam delivered by this source is pre-accelerated in the cyclotron JULIC and injected by charge exchange into the COSY ring. The main diagnostic tool to develop polarized beams in COSY is the EDDA detector [5], primarily designed to measure the pp-scattering excitation function during synchrotron acceleration. The polarization is determined by measuring the asymmetry of scattering between the circulating COSY beam and carbon or CH_2-fiber targets. Additional polarimeters are installed in the injection beamline to COSY, in the COSY ring and in the extraction beamline of COSY. The intensity of the polarized beam in COSY can be increased by stacking injection with an electron-cooler and the beam quality can be further improved with an stochastic cooling systems [6, 7]. The layout of the accelerator complex COSY is shown in Fig. 1.

For an ideal planar circular accelerator with a vertical guide field, the particle spin vector precesses around the vertical axis. Thus the vertical beam polarization is preserved. The spin motion in an external electro-magnetic field is governed by the so called Thomas-BMT equation [8], leading to a spin tune of $v_{sp} = \gamma G$, which describes the number of spin precessions of the central beam per revolution in the ring. G is the anomalous

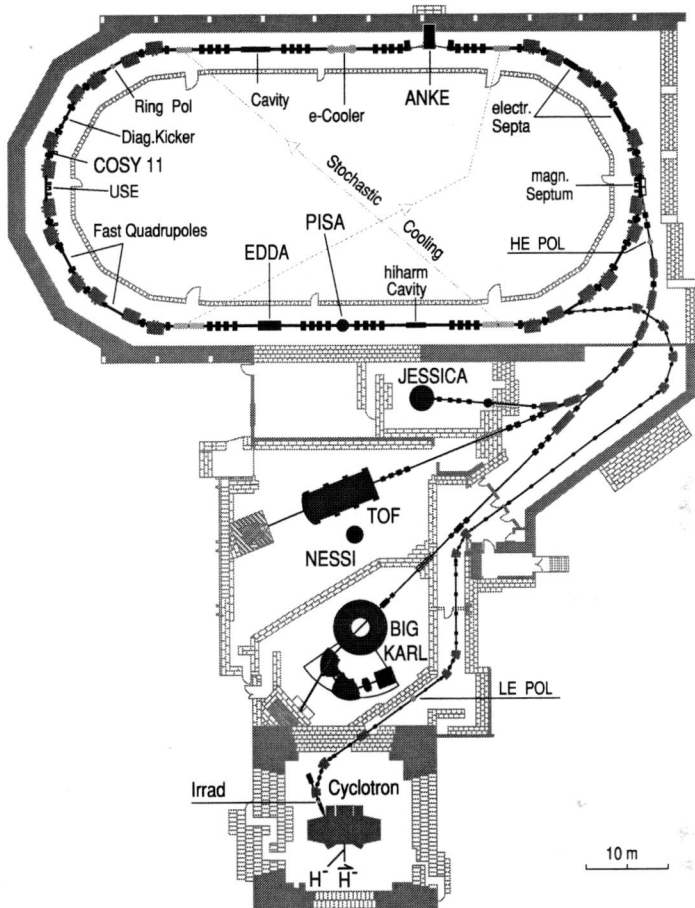

FIGURE 1. The layout of the existing accelerator complex COSY in Jülich, which includes polarized and unpolarized sources, the Cyclotron JULIC and the Cooler Synchrotron COSY. The position of the three different polarimeters (Low Energy Polarimeter, High Energy Polarimeter, Ring Polarimeter) and the four internal experiments (ANKE, COSY 11, EDDA, PISA) are indicated. The beam is also delivered to four external experiment areas (Big Karl, JESSICA, NESSI, TOF).

magnetic moment of the particle (e.g. $G = 1.7928$ for protons), and $\gamma = E/m$ the Lorentz factor. During acceleration of a polarized beam, depolarizing resonances are crossed if the precession frequency of the spin γG is equal to the frequency of the encountered spin-perturbing magnetic fields. In a strong-focusing synchrotron like COSY two different types of strong depolarizing resonances are excited, namely imperfection resonances caused by magnetic field errors and misalignments of the magnets, and intrinsic resonances excited by horizontal fields due to the vertical focusing. For the acceleration of polarized deuterons, additional correction provisions are not necessary to preserve polarization during acceleration, because depolarizing resonances are not crossed in the momentum range of COSY at ordinary transversal betatron tunes [9].

ACCELERATION OF POLARIZED PROTON BEAM AT COSY

In the momentum range of COSY, five imperfection resonances have to be crossed. The existing correction dipoles of COSY are utilized to overcome all imperfection resonances by exciting adiabatic spin flips without polarization losses. The number of intrinsic resonances depends on the superperiodicity of the lattice. In principle the magnetic structure in the arcs of COSY allows to adjust a superperiodicity of P=2 or 6. A tune-jump system consisting of two fast quadrupoles has especially been developed to handle intrinsic resonances at COSY.

Imperfection Resonances

The imperfection resonances for polarized protons are listed in Table 1. They are crossed during acceleration, if the number of spin precessions per revolution of the particles in the ring is an integer ($\gamma G = k$, k: integer). The resonance strength depends on the vertical closed orbit deviation.

TABLE 1. Resonance strength ε_r and the ratio of preserved polarization P_f/P_i at imperfection resonances for a typical vertical orbit deviation y_{co}^{rms}, without considering synchrotron oscillation.

γG	E_{kin} MeV	P MeV/c	y_{co}^{rms} mm	ε_r 10^{-3}	P_f/P_i
2	108.4	463.8	2.3	0.95	-1.00
3	631.8	1258.7	1.8	0.61	-0.88
4	1155.1	1871.2	1.6	0.96	-1.00
5	1678.5	2442.6	1.6	0.90	-1.00
6	2201.8	2996.4	1.4	0.46	-0.58

A spin flip occurs at all resonances without considering synchrotron oscillation. However, the influence of synchrotron oscillation during resonance crossing cannot be neglected (Fig. 2). At the first imperfection resonance, the calculated polarization with a momentum spread of $\Delta p/p = 1 \cdot 10^{-3}$ and a synchrotron tune of $v_{syn} = 6.3 \cdot 10^{-4}$ is about $P_f/P_i{}^1 \approx -0.85$. The resonance strength of the first imperfection resonance has to be enhanced to $\varepsilon_r = 1.6 \cdot 10^{-3}$ for a beam with momentum spread of $\Delta p/p = 1 \cdot 10^{-3}$ to excite spin flips with polarization losses of less than 1%. At the other imperfection resonances the effect of the synchrotron oscillation is smaller, due to the lower momentum

---

[1] Ratio of beam polarization before (i) and after (f) crossing a depolarizing resonance.

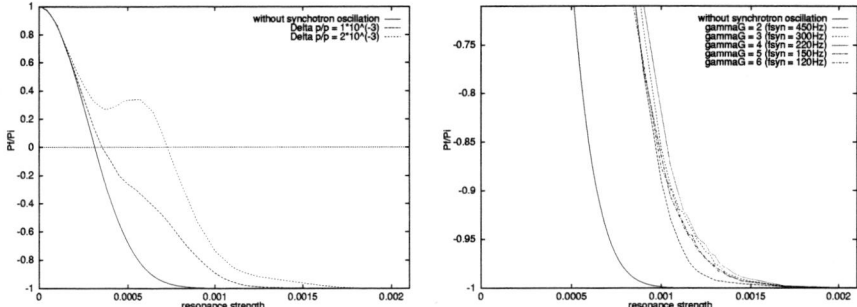

FIGURE 2. Effect of synchrotron oscillation during crossing imperfection resonances in COSY. Ratio of preserved beam polarization P_f/P_i after crossing the first imperfection resonance for two different momentum spreads of $\Delta p/p = 1 \cdot 10^{-3}$ and $\Delta p/p = 2 \cdot 10^{-3}$ with a synchrotron frequency of $\nu_{syn} = 6.3 \cdot 10^{-4}$ (left), and ratio of preserved beam polarization (cutaway in the spin flipping region) after crossing different imperfection resonances for a momentum spread of $\Delta p/p = 1 \cdot 10^{-3}$ taking the synchrotron frequencies at the various resonance energies into account (right). The corresponding synchrotron tunes are in the range between $\nu_{syn} = 6.3 \cdot 10^{-4}$ and $7.7 \cdot 10^{-5}$.

spread at higher energies. Vertical correction dipoles or a partial snake can be used to preserve polarization at imperfection resonances by exciting adiabatic spin flips. Simulations indicate that an excitation of the vertical orbit with existing correction dipoles by 1 mrad is sufficient to adiabatically flip the spin at all imperfection resonances. In addition, the solenoids of the electron-cooler system inside COSY are available for use as a partial snake. They are able to rotate the spin around the longitudinal axis by about 8° at the maximum momentum of COSY. A rotation angle of less than 1° of the spin around the longitudinal axis already leads to a spin flip without polarization losses at all five imperfection resonances [10].

Intrinsic Resonances

The number of intrinsic resonances depends on the superperiodicity P of the lattice, which is given by the number of identical periods in the accelerator. COSY is a synchrotron with a racetrack design consisting of two 180° arc sections connected by straight sections. The straight sections can be tuned as telescopes with 1:1 imaging, giving a 2π betatron phase advance. In this case the straight sections are optically transparent and only the arcs contribute to the strength of intrinsic resonances. One then obtains for the resonance condition $\gamma G = k \cdot P \pm (Q_y - 2)$, where k is an integer and Q_y is the vertical betatron tune. The magnetic structure in the arcs allows adjustment of the superperiodicity to $P = 2$ or 6. The corresponding intrinsic resonances in the momentum range of COSY are listed in Table 2.

TABLE 2. Resonance strength ε_r of intrinsic resonances for a normalized emittance of 1π mm mrad and a vertical betatron tune of $Q_y = 3.61$ for different superperiodicities P.

P	γG	E_{kin} MeV	P MeV/c	ε_r 10^{-3}
2	$6 - Q_y$	312.4	826.9	0.26
2	$0 + Q_y$	950.7	1639.3	0.21
2,6	$8 - Q_y$	1358.8	2096.5	1.57
2	$2 + Q_y$	1997.1	2781.2	0.53
2	$10 - Q_y$	2405.2	3208.9	0.25

Tune-Jump System

A tune-jump system was developed to preserve polarization at intrinsic resonances by increasing the crossing speed significantly. This is accomplished by abruptly changing the vertical betatron tune during resonance crossing in the range of microseconds. The magnet system consists of two pulsed air core quadrupoles and is designed to achieve polarization losses of less than 5% at the strongest intrinsic resonance, and of less than 1% at all other intrinsic resonances in COSY [11]. To meet this goal, a vertical tune jump of more than $\Delta Q_y = 0.06$ in $10\mu s$ is needed. The existing stainless steel vacuum chamber at the location of the tune-jumping quadrupoles had to be replaced by a ceramic vacuum chamber. A layer of $10\mu m$ titanium was sputtered on the inside surface of the ceramic chamber. To avoid double crossing of resonances, the fall time of the tune jump can be adjusted for different jump widths and acceleration rates. The maximum fall time is $40\,ms$. Fig. 3 shows the polarization of the COSY beam measured during acceleration

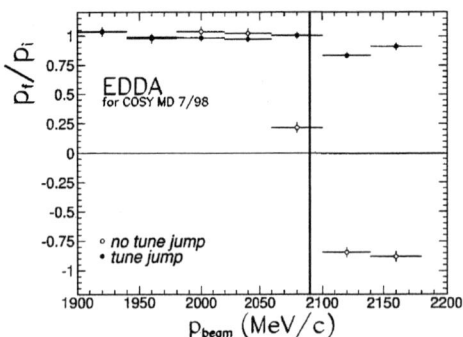

FIGURE 3. Ratio of preserved beam polarization P_f/P_i after crossing the strongest intrinsic resonance at 2090 MeV/c with and without tune jump measured during acceleration with the EDDA detector.

around the strongest intrinsic resonance $\gamma G = 8 - Q_y$. This resonance excites a natural spin flip. The polarization loss depends on the vertical emittance of the beam. With a tune jump, the polarization was almost preserved. Particle losses during tune jumping due to emittance increase can be kept low by adjusting to beam orbit carefully at the position of the tune-jump quadrupoles. The tune jump method can be extended to all other intrinsic resonances because they are at least a factor three weaker than the strongest resonance.

Optimized Optics

To optimize the optics for a polarized beam, phase advances and betatron amplitudes have been determined along the ring. The measurements were done by exciting continuous betatron oscillations and observing the beam response with a network analyzer between a pair of beam position monitors. With the phase advance of the straight sections matched to 2π, the superperiodicity of the COSY lattice is determined by the arcs. Both arcs are composed of three unit cells that are each mirror-symmetrical. A half-cell has a QD-bend-QF-bend structure (Fig. 1). The superperiodicity equals six if all unit cells operate with the same quadrupole settings. In this case only one intrinsic resonance occurs ($\gamma G = 8 - Q_y$) but the transition crossing takes place at about 1600 MeV/c. To accelerate the beam to maximum momentum, the strength of the horizontally focusing quadrupoles of the inner unit cells in the arcs is enhanced by about 40% to shift the transition energy above the maximum momentum. At the same time, the strength of the horizontally focusing quadrupoles in the outer unit cells is decreased by 20% to keep the betatron tunes constant. The superperiodicity of the beam optics is then $P = 2$. Consequently, four additional intrinsic resonances are introduced (Table 2), which can be suppressed if the harmonics of the corresponding spin-perturbing fields are corrected. Theoretical studies of the COSY lattice revealed the possibility of suppressing the strength of intrinsic resonances using the vertically focusing quadrupoles of the inner unit cells in the arcs, leading to a modified P=2-optics [13]. This new method avoids the drawbacks associated with the non-adiabatic nature of tune jumps, which otherwise would be necessary to preserve polarization at all intrinsic resonances of COSY. The method is called suppression of intrinsic spin harmonics, and can also be used at other accelerators like the Brookhaven AGS [12].

During a running period in the year 1998, the new method to overcome intrinsic resonances has been confirmed by measurements with polarized beam (Fig. 4). The polarization was preserved at the two intrinsic resonances, $\gamma G = 6 - Q_y$ and $\gamma G = 0 + Q_y$, by modifying the optics during acceleration. To avoid polarization losses at the first intrinsic resonance, $\gamma G = 6 - Q_y$ at 827 MeV/c, the acceleration of the beam started with P=6 optics. The ratio of preserved polarization was $P_f/P_i = 0.97 \pm 0.05$. At about 900 MeV/c, the COSY beam optics was then switched to superperiodicity P=2 to shift the transition energy. As expected, crossing $\gamma G = 0 + Q_y$ at 1640 MeV/c led to polarization losses ($P_f/P_i = 0.13 \pm 0.05$) in this mode. After suppressing the strength of intrinsic resonances using the vertically focusing quadrupoles in the inner unit cells, the ratio of the preserved polarization at this intrinsic resonance could be significantly increased to $P_f/P_i = 0.88 \pm 0.05$ [13].

However, due to symmetry-breaking installations in the COSY ring (e.g. ANKE spectrometer and electron-cooler magnets) the superperiod of the accelerator lattice

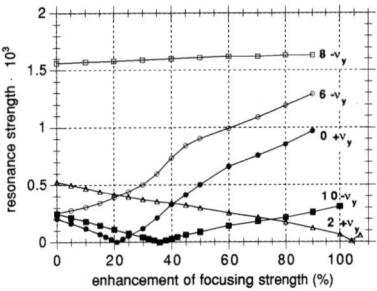 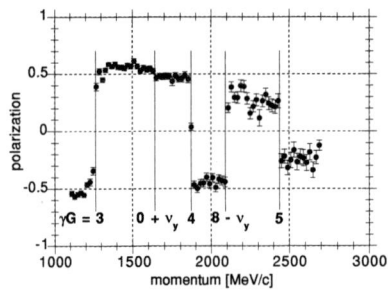

FIGURE 4. The graph on the left side shows the resonance strength of depolarizing resonances in case of a modified $P=2$-optics versus the enhancement of focusing strength of the vertically focusing quadrupoles in the inner unit cells. The betatron tune is fixed by reducing the strength of the vertically focusing quadrupoles in the outer unit cells. In this calculation the focusing strength of the horizontally focusing quadrupoles in the inner unit cells is enhanced by about 40%. The graph on the right side shows the polarization during acceleration measured with the EDDA detector in the momentum range between 1.1 GeV/c and 2.7 GeV/c. The spin was flipped at the imperfection resonances $\gamma G = 3$, 4 and 5. At the second intrinsic resonance $\gamma G = 0 + Q_y$ the polarization was almost preserved by adjusting a modified $P=2$-optics. The third intrinsic resonance $\gamma G = 8 - Q_y$ excites a natural spin flip with some polarization losses.

in COSY is reduced to $P = 1$, leading to five additional intrinsic resonances in the energy range of COSY: ($\gamma G = -1 + Q_y, 7 - Q_y, 1 + Q_y, 9 - Q_y, 3 + Q_y$). To preserve polarization up to maximum momentum of COSY tune jumps are utilized at all ten intrinsic resonances.

Beam Set-up for Polarized Proton Beam

The machine is usually set-up with unpolarized beam, due to better accuracy of COSY diagnostics with about a factor of ten higher beam intensity [14]. Polarization optimization becomes much more efficient and particle losses due to emittance growth can be kept low if the beam position is aligned carefully during the acceleration ramp, especially at the location of the tune-jump quadrupoles. Dynamic tune measurements are carried out to adjust the transversal betatron tunes during acceleration [15]. The vertical betatron tune is fixed close to 3.62 during acceleration in order to optimize the distance between intrinsic resonances for consecutive tune jumps. The horizontal tune is set at around 3.60 during acceleration. After closed orbit and betatron tune correction, the beam manipulations for polarized beam are applied. The magnet currents and trigger times for the tune-jump quadrupoles and vertical correction dipoles are set to values used at previous polarized beam times, as can be seen in Fig. 5. The tuning of magnet currents and trigger times is done after switching to polarized beam by

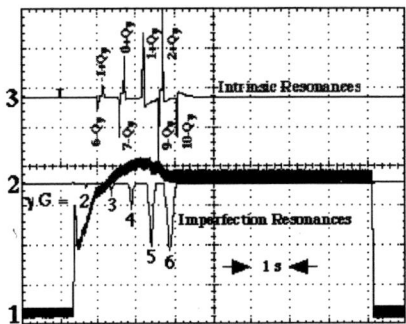

FIGURE 5. Trace 1 shows the beam current, trace 2 the current of vertical correction dipoles, and trace 3 the current of the tune-jump system versus time, applied at various depolarizing resonances.

utilizing polarization measurements. For any applied correction to preserve polarization, the number of particles is observed. Particle losses due to correction dipole and tune-jump quadrupole fields during depolarizing resonance crossing are kept below 10% in total. For extracted polarized beams the momentum has to be chosen carefully to avoid polarization losses. The beam is extracted via a third-order betatron resonance. Corresponding intrinsic resonances lead to significant polarization losses. Momenta near imperfection resonances can also not be provided. Since the beam is stored for relatively long times at extraction energy and because the momentum spread increases during the extraction process, higher order depolarizing resonances can also lead to polarization losses. After excluding these momenta, one still has to carefully adjust the tunes to prepare the stored beam for extraction.

Polarized Proton Beam Acceleration at COSY

During a running period in the year 2000, the polarized beam was accelerated to 3300 MeV/c [14]. The spin was flipped at the imperfection resonances $\gamma G = 2, 3, 4, 5$ and 6 using correction dipoles. To avoid polarization losses at all intrinsic resonances tune jumps were applied. The measured polarization after the optimization for polarized beam is shown in Fig. 6. The polarization losses up to final momentum were rather small, only in the order of a few percent.

Recently, some polarization losses have been observed at the strongest intrinsic resonance $\gamma G = 8 - Q_y$. Further investigations showed, that the coupling resonance $\gamma G = 8 - Q_x$ caused the polarization losses. By moving the horizontal and vertical tunes on top of each other, the polarization after crossing these resonances was almost completely lost. By separating the two transversal tunes, the ratio of preserved polarization increased. A tune separation of $\|Q_x - Q_y\| > 0.15$ was sufficient to reduce the polarization losses at these two resonances to a few percent.

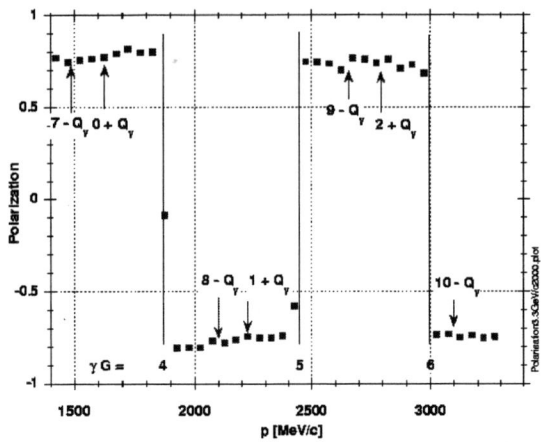

FIGURE 6. Vertical beam polarization during acceleration measured with the EDDA detector in the momentum range between 1100 MeV/c and 3300 MeV/c.

CONCLUSION AND OUTLOOK

The solenoids of the electron-cooler acting as a partial snake and vertical correction dipoles were successfully utilized in COSY to preserve the polarization by exciting adiabatic spin flips. Both methods are available for all five imperfection resonances in the momentum range of COSY. Since solenoids introduce transversal coupling, which excites depolarizing coupling resonances, the vertical correction dipoles are preferred to overcome imperfection resonances in COSY. With the standard optics of COSY, five intrinsic resonances are excited. Measurements confirm, that three of these resonances can be suppressed by changing the optics during acceleration. Due to symmetry-breaking installations, the superperiodicity of the COSY lattice is reduced to one, leading to ten intrinsic resonances. It has been proved that the tune-jump system can handle all ten intrinsic resonances in the momentum range of COSY. Polarization measurements during acceleration confirm that the developed concept allows the acceleration of a vertically polarized proton beam with polarization losses of only a few percent up to the maximum momentum of COSY. The polarization losses at individual depolarizing resonances are within the accuracy of the polarization measurement. Highly polarized proton beams are routinely delivered to internal and external experiments at different momenta. The first injection and acceleration of polarized deuterons is scheduled for February 2003.

REFERENCES

1. Maier R.,*Nucl. Instrum. and Methods* **A 390**, 1 (1997).
2. Stockhorst H. et al., Progress and Developments at The Cooler Synchrotron COSY, Proc. European Particle Accelerator Conference EPAC 2002, Paris, pp. 629-631 (2002).
3. Eversheim P.D. et al., The Polarized Ion-Source for COSY, Proc. International Symposium on High-Energy Spin Physics SPIN 1996, Amsterdam, World Scientific Singapore, pp. 306-308 (1997).
4. Gebel R., Forschungszentrum Jülich, private communication.
5. Schwarz V. et al., EDDA As Internal High-Energy Polarimeter, Proc. International Spin Physics Symposium SPIN 1998, Protvino, World Scientific Singapore, pp. 560-562 (1999).
6. Prasuhn D., Dietrich J., Maier R., Stassen R., Stein H.J., Stockhorst H., *Nucl. Inst. and Meth.* **A 441**, 167 (2000).
7. Prasuhn D. et al., Cooling at COSY, Workshop on Beam Cooling and Related Topics, Bad Honnef 2001, Forschungszentrum Jülich, Matter and Material, Volume **13**, ISBN 3-89336-316-5.
8. Thomas L.H. *Phil. Mag.* **3**, 1 (1927); Bargman V., Michel L., Telegdi V.L., *Phys. Rev. Letters* **2**, 43 (1959).
9. Lehrach A. et al., Acceleration of Polarized Protons and Deuterons at COSY, Proc. International Spin Physics Symposium SPIN 2002, Brookhaven, to be published in AIP Conference Proceedings.
10. Lehrach A. et al., Status of the polarized beam at COSY, Proc. International Symposium on High-Energy Spin Physics SPIN 1996, Amsterdam, World Scientific Singapore, pp. 416-418 (1997).
11. Lehrach A., PhD-thesis Universität Bonn (1997), Jülich Report Juel-3501, ISSN 0944-2952 (1998).
12. Lehrach A., Ranjbar, V.H., AGS Lattice Changes to eliminate Weak Intrinsic Resonances, this conference proceedings.
13. Lehrach A., Gebel R., Maier R., Prasuhn D., Stockhorst H., *Nucl. Instrum. and Methods* **A 439**, 26 (2000).
14. Stockhorst H. et al., The Medium Energy Proton Synchrotron COSY, Proc. European Particle Accelerator Conference EPAC 2000, Vienna, pp. 590-592 (2000).
15. Dietrich J., Mohos I., Broadband FFT Method for Betatron Tune Measurements in the Acceleration Ramp at COSY-Jülich, 8th Beam Instrumentation Workshop, Stanford, AIP Conference Proc. No. 451, 454 (1998).

20% Partial Siberian Snake in the AGS[1]

H. Huang, L. Ahrens, M. Bai, K.A. Brown, W. Glenn, A.U. Luccio,
W.W. MacKay, C. Montag, V. Ptitsyn, T. Roser, N. Tsoupas, K. Zeno,*
V. Ranjbar,† H. Spinka and D. Underwood **

C-A Department, Brookhaven National Laboratory, Upton, NY 11973, USA
†*Physics Department, Indiana University, Bloomington, IN 47405*
**Argonne National Laboratory, Argonne, IL 60439*

Abstract. An 11.4% partial Siberian snake was used to successfully accelerate polarized protons through a strong intrinsic depolarizing spin resonance in the AGS. No noticeable depolarization was observed. This opens up the possibility of using a 20% to 30% partial Siberian snake in the AGS to overcome all weak and strong depolarizing spin resonances. Some design and operation issues of the new partial Siberian snake are discussed.

AGS POLARIZED PROTON RUNS SINCE 1990'S

Polarized protons have been accelerated in the Brookhaven Alternating Gradient Synchrotron (AGS) since 1980s. In 1980s, it was used for fixed target high P_T proton-proton elastic scattering experiment[1]-[2]. Since 1990s, AGS polarized proton program has been running to develop the AGS as injector for spin physics program in RHIC[3]. The goal for the AGS is to deliver 2×10^{11} proton/bunch with 70% polarization at 24GeV/c.

Acceleration of polarized proton beams to high energy in circular accelerators is difficult due to numerous depolarizing resonances. During acceleration, a depolarizing resonance is crossed whenever the spin precession frequency equals the frequency with which spin-perturbing magnetic fields are encountered. In the presence of the vertical dipole guide field in an accelerator, the spin precesses $G\gamma$ times per orbit revolution [4], where $G = (g-2)/2 = 1.7928$ is the coefficient of the gyromagnetic anomaly of the proton, and γ is the Lorentz factor. The number of precessions per revolution is called the spin tune v_{sp} and is equal to $G\gamma$ in this case.

There are three main types of depolarizing resonances in the AGS: imperfection resonances, which are driven by magnet misalignments; intrinsic resonances, driven by the vertical betatron motion through quadrupoles; and coupling resonances, caused by the vertical motion with horizontal betatron frequency due to linear coupling [5]. The resonance condition for an imperfection resonance is $v_{sp} = n$, where n is an integer. The resonance condition for an intrinsic resonance is $v_{sp} = nP \pm v_y$, where n is an integer, $P = 12$ is the superperiodicity of the AGS, and v_y is the vertical betatron tune. The resonance condition for a coupling spin resonance is $v_{sp} = n \pm v_x$, where v_x is

---

[1] This work was supported by the Department of Energy of United States.

the horizontal betatron tune; it is only important if v_{sp} is close to a strong intrinsic resonance condition. The spin resonance strength ε_k is defined as the Fourier amplitude of the spin perturbing field. In general, the intrinsic resonance strength is proportional to $\sqrt{\gamma \varepsilon_N}$, where ε_N is the normalized vertical emittance of the beam; and the imperfection resonance strength is proportional to γ for a given closed orbit error. When a polarized beam is uniformly accelerated through an isolated spin resonance, the final polarization P_f is related to the initial polarization P_i by the Froissart-Stora formula[6]

$$P_f = (2e^{-\pi|\varepsilon_k|^2/2\alpha} - 1)P_i, \qquad (1)$$

where α is the resonance crossing rate given by $\alpha = \frac{d(G\gamma)}{d\theta}$, and θ is the orbital bend angle in the synchrotron.

In the early days, the polarization was preserved by non-adiabatic techniques, tune-jump for intrinsic resonances and harmonic orbit correction for imperfection resonances[1]-[2]. Over the years, the adiabatic techniques have prevailed. Several novel schemes have been developed to overcome these resonances in the AGS. A 5% partial solenoidal Siberian snake [7] has been used successfully to overcome imperfection resonances [8].

For a ring with a partial Siberian snake of strength s, the spin tune v_{sp} is given by [9]

$$\cos \pi v_{sp} = \cos \frac{s\pi}{2} \cos G\gamma \pi, \qquad (2)$$

where $s = 1$ corresponds to a full Siberian snake which rotates the spin by 180°. When s is small, the spin tune is nearly equal to $G\gamma$ except when $G\gamma$ equals an integer n, where the spin tune v_{sp} is shifted away from the integer by $\pm s/2$. Since the spin tune never equals an integer, the imperfection resonance condition is never satisfied. Thus the partial Siberian snake can overcome all imperfection resonances, provided that the resonance strengths are much smaller than the spin tune gap created by the partial Siberian snake. For the AGS, nominally $\alpha = 4.8 \times 10^{-5}$, resonance strength $|\varepsilon_k| < 0.01$ for all imperfection resonances from experience of earlier runs [1]-[2], a 5% partial Siberian snake is enough to overcome all imperfection resonances.

As shown in Fig. 1, the measured asymmetry changed sign at every $G\gamma$= integer because of the partial Siberian snake. The polarization loss at $G\gamma = 42.3$ is caused by a hybrid resonance [10]. The additional structure between integers is due to the numerous weak resonances at $G\gamma = k \pm v_y$ where integer k is not an integer multiple of superperiodicity of AGS. Since these resonance strength are very weak, it would not cause noticeable polarization loss at nominal resonance crossing rate. However, the asymmetry for each energy of Fig. 1 was measured by coasting beam at that energy. Therefore, the resonance crossing rate was very low and the depolarizing effect at these resonances was enhanced. More importantly, within error bars, the polarization measured at half integers agrees with the expected values, except $G\gamma = 39.5$, where intrinsic resonance $G\gamma = 48 - v_y$ located. This confirmed that a 5% partial Siberian snake is strong enough to overcome imperfection resonances up to high energy in the AGS and full spin flip was achieved at all imperfection resonances.

By adiabatically exciting a vertical coherent betatron oscillation using a single ac dipole magnet, an artificial spin resonance is excited. If the resonance location is chosen

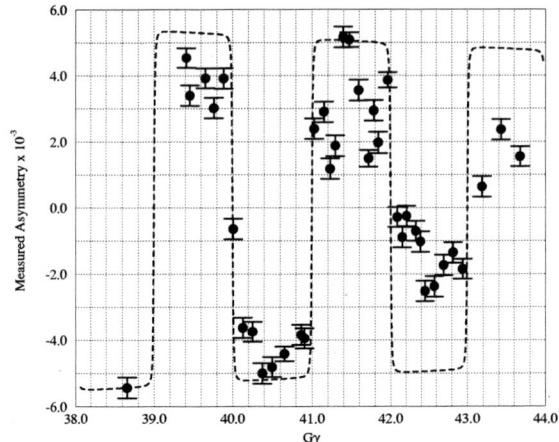

FIGURE 1. The measured asymmetry for different beam energies or $G\gamma$ values. The dashed line shows the asymmetry expected from a 5% partial Siberian snake. It is scaled based on the assumption that the analyzing power is inversely proportional to the beam momentum.

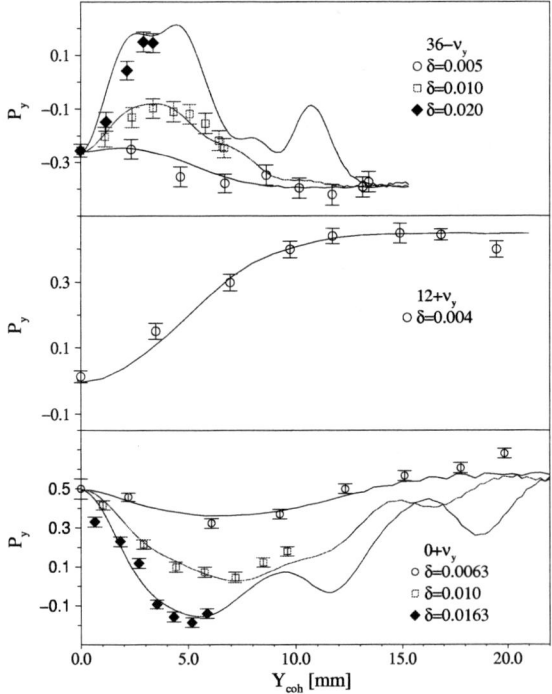

FIGURE 2. Measured polarization as function of generated coherent betatron motion amplitude Y_{coh} for various tune separations with an ac dipole. Note that when driving amplitude is large enough, all curves reach plateaus, which indicate full spin-flip is achieved.

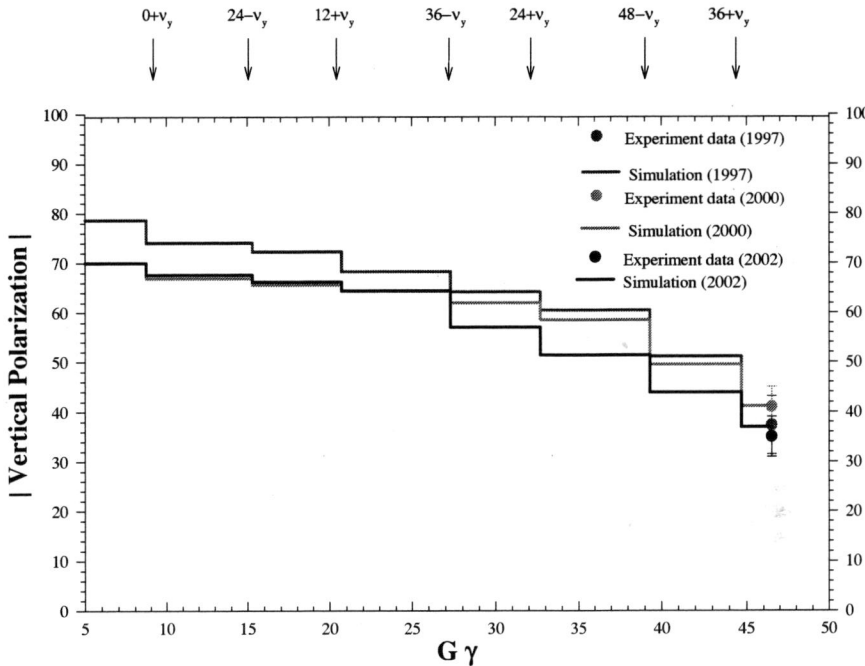

FIGURE 3. Summary of polarization in the AGS. The solid lines are the SPINK simulation results.

near an intrinsic spin resonance, the spin motion will be dominated by the ac dipole resonance, and full spin-flip can be achieved without significant emittance growth [11]. Fig. 2 shows the measured polarization as function of ac dipole driving amplitude and tune separation δ (difference between the modulation tune and the vertical betatron tune). It reveals that we have achieved full spin-flip at $0+v_y$, $12+v_y$ and $36-v_y$ within the uncertainty of the AGS internal polarimeter. It should be noted that no such data exists for $36+v_y$. At high energies, the existing polarimeter has small analyzing power and polarization level is lower, so it is very difficult to make such a plot for $36+v_y$.

Due to the linear coupling induced by the solenoidal field of the partial snake, coupling depolarizing resonances are enhanced. The two betatron tunes have to be well separated to reduce the coupling effect[5]. The typical betatron tune separation is 0.15.

In summary, the proton polarization in the AGS is shown in Fig. 3. Simulations done with SPINK [5] are also included in Fig. 3. Full spin-flip is achieved at all imperfection resonances using the 5% partial Siberian snake, and at all strong intrinsic resonances using an ac dipole. The remaining polarization loss comes from coupling and weak intrinsic resonances. The steps in the polarization levels are the polarization losses due to coupling resonances and weak intrinsic resonances.

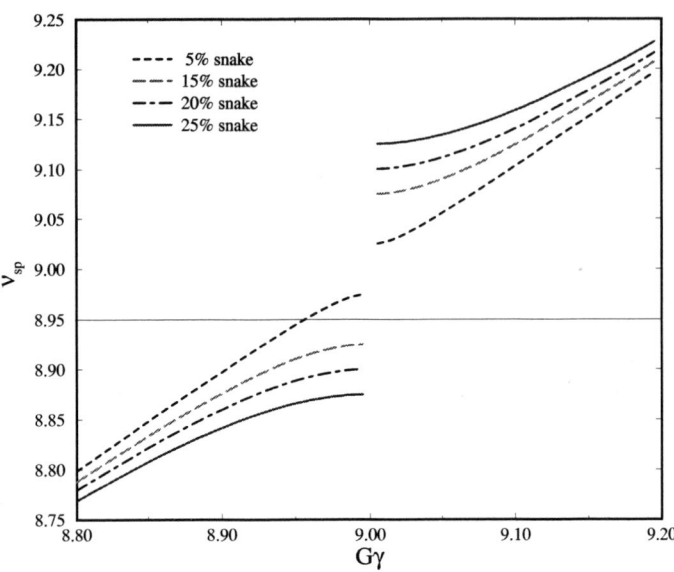

FIGURE 4. Spin tune for various partial Siberian snake strengths. The straight line indicates a possible value for the vertical betatron tune.

WHY STRONG PARTIAL SNAKE?

As shown in Fig. 3, the remaining depolarization in the AGS is due to the weak intrinsic resonances and coupling resonances which are enhanced by the presence of the solenoidal partial Siberian snake. To reduce coupling resonance strength, a new snake with less coupling is needed. The ac dipole technique only works for strong intrinsic resonances, since it relies on the strength of the intrinsic resonances to induce a strong enough artificial resonance. In addition, ac dipole operation need precise control of betatron tunes to the level of 0.0005 and need to be constantly monitored for polarized proton operation. Currently, there is no scheme to overcome the weak intrinsic resonances. It should be noted that two other techniques presented in this workshop may also work for the weak intrinsic resonances: tune-jump [12] and spin harmonic matching [13].

As can be seen from Fig. 4, with a strong enough partial Siberian snake, the spin tune gap can be increased to allow placing the betatron tune inside the gap so that the intrinsic resonance conditions can also be avoided. This idea has been proposed previously in Refs. [14] and [15]. Simulations showed that for the first intrinsic resonance at $0+v_y$, a 10% partial Siberian snake would be strong enough. In a recent experiment at the AGS this was successfully demonstrated.

For the AGS, this method has several advantages. First, it works for both strong and weak intrinsic resonances. Currently, there is no effective way to overcome the weak intrinsic resonances in the AGS. Second, if the coupling of a new Siberian snake could be reduced, the strength of coupling resonances could also be reduced. Or, if both horizontal and vertical betatron tunes could be put into the spin tune gap, both intrinsic and coupling resonances could be avoided.

STRONG SNAKE EXPERIMENT

The polarized H$^-$ beam from the optically pumped polarized ion source [16] was accelerated through a radio frequency quadrupole and the 200 MeV LINAC. The beam polarization at 200 MeV was measured with elastic scattering from a carbon fiber target. A fast switching magnet assured the polarization can be measured on subsequent pulses at the end of the LINAC and in the AGS. During the study, the polarization measured by the 200 MeV polarimeter was (66±0.5)%. The beam was then strip-injected and accelerated in the AGS Booster up to 1.5 GeV kinetic energy ($G\gamma = 4.7$). The vertical betatron tune of the AGS Booster was chosen to be 4.9 in order to avoid crossing the intrinsic resonance at $G\gamma = 0 + v_y$ in the Booster. The imperfection resonances at $G\gamma = 3$, 4 in the Booster were corrected by harmonic orbit correctors.

Only one bunch of the twelve rf buckets in the AGS was filled, and the beam intensity varied between 1.3 -1.7×10^{11} protons per fill. The polarized proton beam was accelerated up to 5.6 GeV kinetic energy ($G\gamma = 12.5$) passing through just one intrinsic resonance located at $G\gamma = 0 + v_y$. The resonance crossing rate α was 2.4×10^{-5}. Polarization was measured at $G\gamma = 12.5$ during an approximately one second flattop after the partial Siberian snake was ramped to zero. The spin rotation angle ϕ in the solenoid is given by

$$\phi = e(1+G)\mu_0 NI/cp, \tag{3}$$

where p is the momentum of the proton beam, μ_0 is the permeability of vacuum, and NI is the current in ampere turns. The effective Siberian snake strength s of the partial Siberian snake is $s = \phi/\pi$. The solenoidal partial Siberian snake in the AGS is capable of achieving a 5% Siberian snake strength at 24.2 GeV kinetic energy ($G\gamma = 48$). At $G\gamma = 0 + v_y$, the solenoid can in principle generate a 25% partial Siberian snake. However, the solenoidal field will also generate significant coupling, which will cause sizeable depolarization. In addition, such a strong Siberian snake will tilt the stable spin direction away from vertical by 12.5°, reducing the measurable vertical polarization component. For this experiment an 11.4% partial Siberian snake was chosen as a compromise between obtaining a large enough spin tune gap and minimizing the coupling effects. The AGS partial Siberian snake was turned on to 6% before beam injection into the AGS and then ramped up to 11.4% before the first intrinsic resonance crossing at $0 + v_y$. The orbit was carefully corrected to maintain beam stability as the vertical betatron tune was moved as high as 8.98. During the experiment, the horizontal betatron tune was kept at 8.54, while the beam polarization was measured as a function of the vertical betatron tune.

RESULTS AND DISCUSSION

The experimental data and simulation results are plotted in Fig.5. The polarization was measured with the AGS internal polarimeter [17]. Measured vertical betatron tunes were

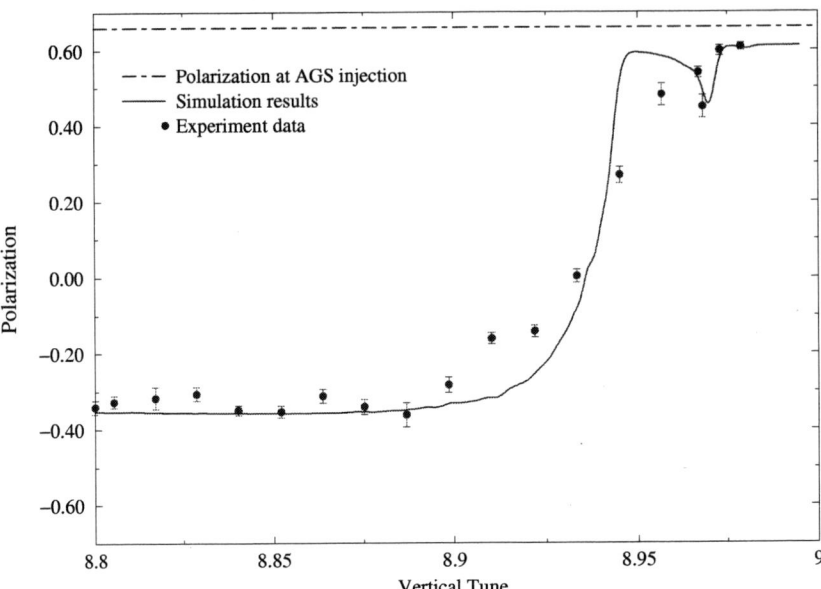

FIGURE 5. The measured vertical polarization as a function of the vertical betatron tune for an 11.4% partial Siberian snake. The dots are measured polarizations, and the error bars represent the statistical errors only. The dashed straight line indicates the polarization level measured at the end of the LINAC. Since the two imperfection resonances in the Booster have been corrected by harmonic orbit correctors, this is also the beam polarization at AGS injection. The solid curve shows the simulation results.

not available in the tune window $v_y = 8.90$ to 8.96. A fit of set tunes to measured tunes outside this window was used to derive the vertical betatron tunes from the set values inside this window. To estimate the depolarizing resonance strength, both vertical and horizontal beam emittance were measured using the AGS ionization profile monitor(IPM) except for the data points in the tune window $v_y = 8.90$ to 8.96. As can be seen from Fig.5, the measured polarization reached a plateau when the vertical betatron tune was very close to nine. The polarization loss in this region was only about 6% and can be completely explained by spin mismatch at AGS injection and depolarization from coupling resonances as discussed below.

These observations agree well with spin dynamics calculations. With a partial Siberian snake inserted, there are two strong resonances in this energy region: one located at $G\gamma = 9$ generated by the partial Siberian snake and the intrinsic resonance at $G\gamma = 0 + v_y$. When the intrinsic and imperfection resonances do not overlap ($v_y \leq 8.85$), the resonance at $G\gamma = 9$ should flip the spin completely while the intrinsic resonance at $G\gamma = 0 + v_y$ causes some depolarization. When the two resonances are very close, such as for $v_y = 8.98$, the intrinsic resonance is overpowered by the resonance at $G\gamma = 9$. The particles essentially just experience one resonance at $G\gamma = 9$, and full spin-flip is observed. When the two resonances are at intermediate separations, such as for $v_y \approx 8.90$ to 8.95, the two resonances interfere with each other.

A simulation was performed to better understand the polarization behavior in this experiment. The simulation is a combination of a DEPOL [18] calculation and a tracking model with two overlapping resonances: one located at $G\gamma = 9$ generated by the partial Siberian snake and the intrinsic resonance at $G\gamma = 0 + v_y$. The strength of the intrinsic resonance was determined from beam size measurements. A ± 0.004 vertical tune spread was included in the simulation.

The strengths of the coupling resonances located at $G\gamma = 17 - v_x, 0 + v_x, 18 - v_x$, and $1 + v_x$ were calculated using an extended DEPOL that was modified to include coupling [19]. Since these coupling resonances are well separated from the other two resonances, they can be treated independently. The total polarization loss from the coupling resonances was calculated, using the Froissart-Stora formula, to be 5% out of the total 6% polarization loss shown in Fig. 5. The remaining 1% loss is due to spin mismatch at injection with a 6% partial Siberian snake.

In general, the simulation agrees well with most data points. The remaining discrepancies for data points between $v_y = 8.90$ to 8.96 could be due to a different beam size or vertical betatron tune, since there were no beam size and tune measurements performed for these data points.

The simulation shows a polarization dip close to $v_y=8.97$, which may also be seen in the experimental data. This is caused by a snake resonance [20] as predicted in Refs. [15] and [21]. Even when the intrinsic resonance condition can not be met for $v_y > 8.943$, depolarization can occur from resonance conditions extended over many turns if the intrinsic resonance is very strong. This happens when the following condition is met

$$\Delta v_y = \frac{k \pm v_{sp}}{n}, \qquad (4)$$

where Δv_y is the fractional part of vertical betatron tune, n and k are integers, and n is called the snake resonance order. With an 11.4% partial Siberian snake, the spin tune is close to 0.057 for $G\gamma \sim 9$. The polarization dip then corresponds to the second order snake resonance ($n = 2$). With the given resonance crossing rate and intrinsic resonance strength, snake resonances higher than second order do not show a significant effect. The existence of this snake resonance reduces the usable betatron tune space where depolarization is avoided.

At the vertical betatron tune of 8.98, the difference between the beam polarization at injection and the polarization measured after $0 + v_y$ is due to spin mismatch at injection and depolarization from coupling resonances. If spin matching were achieved at the AGS injection and the linear coupling were eliminated, this scheme could provide full spin-flip through the intrinsic resonance. It would also work for weak intrinsic resonances, such as $G\gamma = 24 + v_y$, and $48 - v_y$. In addition, if the horizontal betatron tune were also in the gap, the coupling resonance could also be avoided.

SNAKE DESIGN ISSUES

Simulation shows that a 20% partial Siberian snake is needed for the strongest intrinsic resonance at $36 + v_y$. This is beyond the capability of the existing solenoidal partial

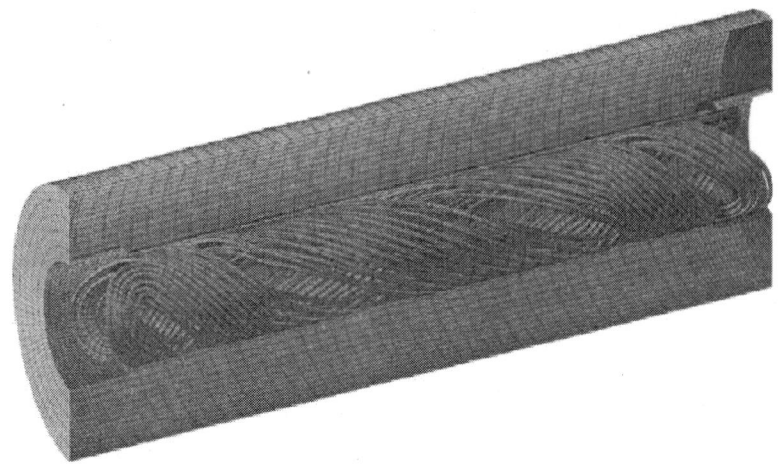

FIGURE 6. The design of superconducting snake magnet. It is 2.6 meters long and the diameter of the beam pipe is 15 cm.

snake. Furthermore, the solenoidal field is the main source of coupling, which causes coupling resonances in the vicinity of strong intrinsic resonances. The better choice would be a helical dipole magnet as has been used in RHIC polarized proton operation. With the constraint of 10-foot AGS straight section, the required field can only be achieved by a super-conducting magnet. With compensating coils, the coupling from the new snake can be greatly reduced. The drawback of a super-conducting magnet is the complicated cryogenic system, which needs special attention to cope with high intensity proton operation in the AGS. For high intensity proton operation, the magnet is power-off but stay cold. The radiation loss heat load at a quiet AGS straight section is estimated from beam loss monitor to be $1 \sim 2$ Watt for high intensity (7×10^{13} protons/spill), which is manageable. For polarized proton operation, the magnet stays power-on and the snake may be the limiting aperture. To prevent catastrophic beam loss in the snake magnet, a collimator upstream will be needed for the polarized proton operation as a protection device.

At injection energy, the maximum orbit excursion inside the helical snake is ± 3 cm in both horizontal and vertical planes. Given 15πmm-mrad normalized emittance at injection, the 15cm diameter beam pipe should be enough. For high intensity proton operation, the snake is off and there is no orbit excursion. No aperture problem exists for the 100πmm-mrad high intensity beam.

The design of a super-conducting helical AGS snake with strength on the order of 20% to replace the current solenoidal AGS partial snake has already begun. A sketch of the design is shown in Fig. 6. With the given field, the snake strength at AGS injection would be about 24%. At injection energy, the tune shifts are of the order of 0.2 units, and the beta functions fluctuate throughout the ring up to values around 100 m instead of the matched maximum of about 22 m. A solution has been found by E. Courant [22] to eliminate the coupling and beta function mismatch caused by the snake.

In addition, such a strong snake in the AGS will tilt the stable spin direction away from vertical significantly. Current design gives 24% partial snake at injection, which will tilt away the stable spin direction (SSD) at the AGS injection from vertical by 21.6°. A straightforward injection gives 7% polarization loss. One possible scheme to match the spin is to raise the injection energy to get $G\gamma$ from 4.7 to 5. Then the SSD is in horizontal plane in both Booster and the AGS. A proper choice of 5th harmonic corrector setting of the Booster can match the SSD in the Booster to the SSD in the AGS. To avoid $0 + \nu_y$ intrinsic resonance in the Booster, the Booster vertical tune should be set at 5.1 (currently at 4.7), which is within the capability of Booster tune quadrupoles. Spin matching should also be done at AGS extraction and a scheme has been reported in Ref. [23].

CONCLUSION

In conclusion, we have demonstrated for the first time that an 11.4% partial Siberian snake can effectively overcome an intrinsic depolarizing resonance when the vertical betatron tune is put close to an integer. The critical element of this operation is to maintain beam stability under these conditions. The challenge will be spin matching at AGS injection and extraction, high order field compensation and cryogenic system operation.

REFERENCES

1. F.Z. Khiari, et al., *Phys. Rev.* **D39**, 45 (1989).
2. L. Ahrens, in *Proceedings of the 8th International Symposium on High-Energy Spin Physics*, Minneapolis, 1988, AIP Conf. Proc. No **187** (AIP, New York, 1989), p.1068.
3. I. Alekseev, et al., *Design Manual of Polarized Proton Collider at RHIC*, 1998, unpublished.
4. L.H. Thomas, *Philos. Mag.* **3**, 1 (1927); V. Bargmann, L. Michel, and V.L. Telegdi, *Phys. Rev. Lett.* **2**, 435 (1959).
5. H. Huang, T. Roser, A. Luccio, *Proc. of 1997 IEEE PAC*, Vancouver, May, 1997, p.2538.
6. M. Froissart and R. Stora, *Nucl. Instrum. Methods* **7**, 297(1960).
7. Ya.S. Derbenev and A.M. Kondratenko, *Part. Accel.* **8**, 115 (1978).
8. H. Huang, et al., *Phys. Rev. Lett.* **73**, 2982 (1994).
9. T. Roser, in *Proceedings of the 8th International Symposium on High-Energy Spin Physics*, Minneapolis, 1988, AIP Conf. Proc. No **187** (AIP, New York, 1989), p.1442.
10. M. Bai, et al., *Phys. Rev. Lett.* **84**, 1184 (2000).
11. M. Bai, et al., *Phys. Rev. Lett.* **80**, 4673 (1998).
12. M. Bai, these proceedings.
13. A. Lehrach, these proceedings.
14. T. Roser, in the Polarized Proton in the AGS Workshop, 1990, unpublished.
15. S.Y. Lee, *Spin Dynamics and Siberian Snakes in Synchrotrons*, World Scientific (1997).
16. A. Zelenski, et al., in *Proceedings of the 9th International Conference on Ion Sources*, Rev. Sci. Inst., Vol.73, No.2, p.888 (2002).
17. C.E. Allgower, et al., *Phys. Rev.* **D65**, 092008(2002).
18. E.D. Courant and R.D. Ruth, BNL report 51270, (1980).
19. V. Ranjbar, et al., *Proc. of 2001 IEEE PAC*, Chicago, June, 2001, p. 3177.
20. S.Y. Lee and S. Tepikian, *Phys. Rev. Lett.* **56**, 1635 (1986).
21. S.Y. Lee, *Phys. Rev.* **E47**, 3631 (1993).
22. E. Courant, these proceedings.
23. W.W. MacKay, these proceedings.

Review of Polarized Proton Beam Acceleration at KEK-PS in the 1980's

C. Ohmori, S. Hiramatsu, H. Sato, and T. Toyama

KEK, Oho 1-1, Tsukuba, Ibaraki 305-0801, Japan

Abstract. A polarized proton beam was accelerated at the KEK-PS in the late 1980's. This report will review the activities for the polarized beam acceleration in both the 500 MeV booster and the 12 GeV Main Ring. Some depolarization resonances were passed using the spin-flip techniques. The effects on the polarization caused by a synchrotron side-band at a strong intrinsic resonance were observed.

INTRODUCTION

In 1980, the polarized beam project was started at the KEK-PS. An optically pumped polarized ion source was developed [1]. The first acceleration test in the 500 MeV booster was performed in 1983 [2]. After a 14-months shut-down of KEK-PS due to the construction of the TRISTAN tunnel, the beam polarization at the booster was increased to 40% in 1985 [3]. In 1989, for a 64% polarized beam at 20 MeV before injection, a maximum polarization of 54% was obtained at 500 MeV [4]. In 1986, the polarized beam was accelerated up to 3.5 GeV in the Main Ring. The beam polarizations were 44% at 500 MeV and 38% at 3.5 GeV, respectively [5,6]. Two nuclear experiments were performed using the polarized proton beam. In 1987, 25% polarization was obtained at 5.0 GeV and about 5% at 7.6 GeV [7, 8]. In this paper, the resonance corrections up to 5.0 GeV will be described.

EQUIPMENT FOR POLARIZED BEAM ACCELERATION

The equipment is shown in Fig. 1. To measure the beam polarization at different stages of acceleration, four polarimeters [9,10,11] were installed: a 20 MeV polarimeter between the two linac tanks, Injection and Main Ring polarimeters in the Main Ring, and an external polarimeter in the KEK-PS counter hall. The last three polarimeters consisted of double-arm counter systems to measure the proton-proton elastic scattering at fixed energies. Polyethylene and carbon targets were used for the subtraction of the carbon background. The Main Ring polarimeter was also used to measure the asymmetry of backward counter telescopes during the acceleration. This was a useful method to identify the timing of the depolarization resonances as shown in Fig. 2. Because of high counting rates, the asymmetry measurements were finished quickly.

In the booster ring, two pulsed quadrupole magnets were installed to pass through the resonances. These two magnets were replaced by two sextupole magnets to correct for the chromaticity. Two pulsed dipole magnets were also installed to correct the closed orbit. In the Main Ring, four pulsed quadrupole magnets were installed to cross the intrinsic resonances. The rise time of the magnets were 40 µsec to 200 µsec. Twenty-eight correction dipole magnets were located near the defocusing quadrupole magnets to correct the closed orbit.

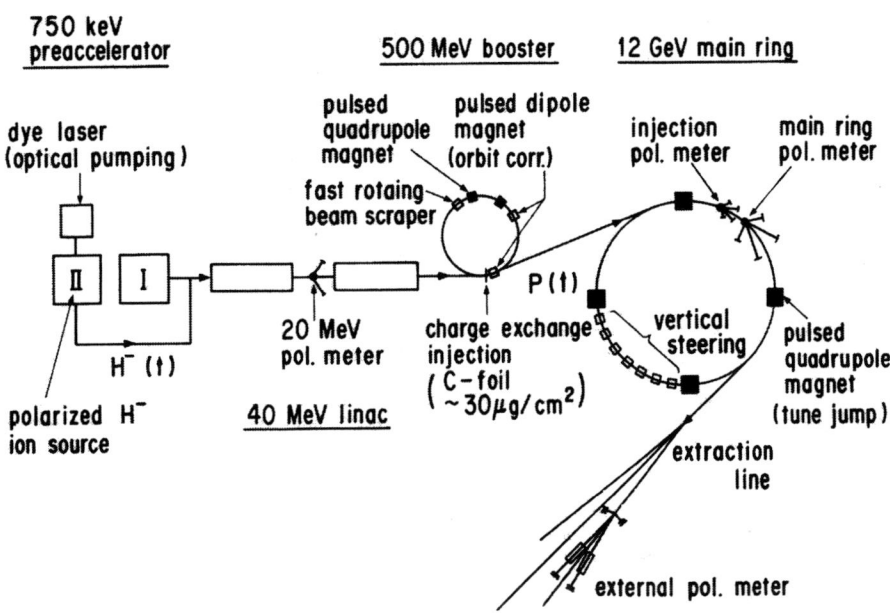

FIGURE 1. Equipment for the polarized beam acceleration.

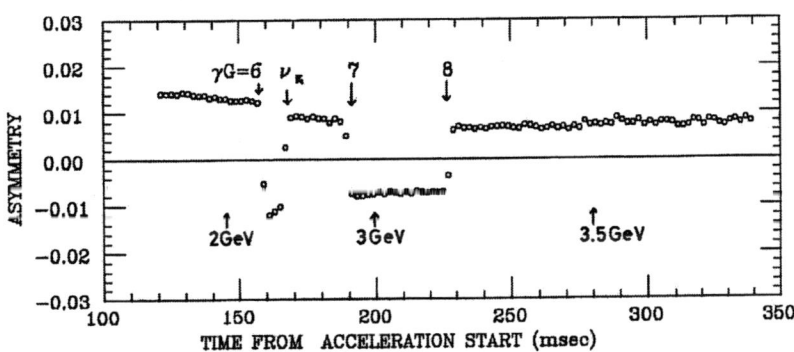

FIGURE 2. Asymmetry measured with a fixed backward counter telescope during the acceleration up to 3.5 GeV.

DEPOLARIZATION RESONANCES AT KEK-PS

The depolarization resonances in the booster ring are listed in Table 1. The booster is a combined-type synchrotron which includes a strong sextupole field component. There are two strong resonances (one intrinsic and one imperfection) and two weak (higher order) resonances. The intrinsic resonance occurs at

$$\gamma G = n N \pm \nu_z, \qquad (1)$$

where n is an integer, N is the super-periodicity of the synchrotron and ν_z is the vertical betatron tune. The imperfection one occurs at

$$\gamma G = n N \pm k, \qquad (2)$$

where k is the harmonic number of the closed orbit distortion.

The Main Ring has a super-periodicity of four and has more resonances than other synchrotrons. The intrinsic resonances are listed in Table 2. It was expected that the resonances $\gamma G = 12 - \nu_z$, $16 - \nu_z$, and $20 - \nu_z$ could cause significant beam depolarization and that the others were strong enough to make adiabatic spin-flips without large depolarizations. The pulsed quadrupole magnets were designed to pass the three medium resonances [12].

TABLE 1. Depolarizing Resonances in the Booster.

γG	T (MeV)	ε
2	108	$2.6 \cdot 10^{-3}$
ν_x	200	$2.2 \cdot 10^{-4}$
ν_z	280	$1.4 \cdot 10^{-2}$
$5-\nu_z$	440	$1.2 \cdot 10^{-4}$

TABLE 2. Intrinsic Resonances below 7.6 GeV in the Main Ring.

γG	T (GeV)	ε
$12-\nu_z$	2.1	$2.0 \cdot 10^{-3}$
ν_z	2.3	$2.2 \cdot 10^{-2}$
$16-\nu_z$	4.2	$2.8 \cdot 10^{-3}$
$4+\nu_z$	4.4	$1.0 \cdot 10^{-2}$
$20-\nu_z$	6.3	$6.4 \cdot 10^{-4}$
$8+\nu_z$	6.5	$1.3 \cdot 10^{-2}$

IMPERFECTION RESONANCES

In the booster, the imperfection resonance, $\gamma G = 2$, was strong enough to excite an adiabatic spin flip using a vertical deflector as shown in Fig. 3.

In the Main Ring, the resonances, $\gamma G = 3, 4, 5$ and 9, were weak and no depolarization was observed. The resonances, $\gamma G = 6$ and 8, were strong and they were passed by exciting the 6th and 8th harmonic component. For the resonances, $\gamma G = 10$

and 11, a rather complicated correction scheme was applied. These resonances were not only affected by the 10th and 11th harmonic components, but also by the 6th and 7th ones respectively, as indicated by equation (2).

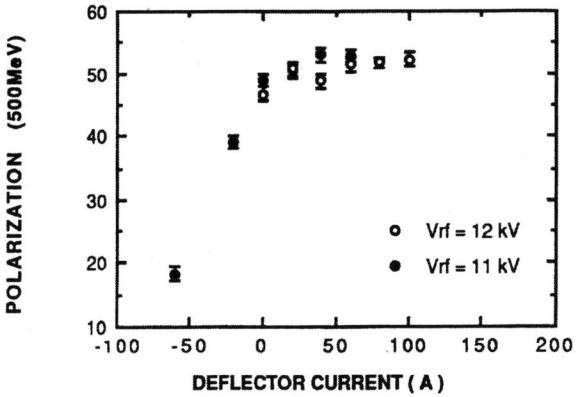

FIGURE 3. Deflector current dependence of the strong imperfection resonance, $\gamma G = 2$.

CORRECTION OF INTRINSIC RESONANCES

In the Main Ring, the two intrinsic resonances, $\gamma G = 12 - v_z$ and $16 - v_z$, were located below 5 GeV and have medium strengths to cause significant depolarizations. These were corrected by the fast passage method using pulsed quadrupole magnets. A typical rise-time to overcome the resonance $\gamma G = 12 - v_z$ was 70 μsec; variation of the vertical tune was 0.15. The rise-time for the resonance $\gamma G = 16 - v_z$ was 140 μsec; variation of the vertical tune was 0.17. Figure 4 shows the correction curves for the intrinsic resonance $\gamma G = 12 - v_z$.

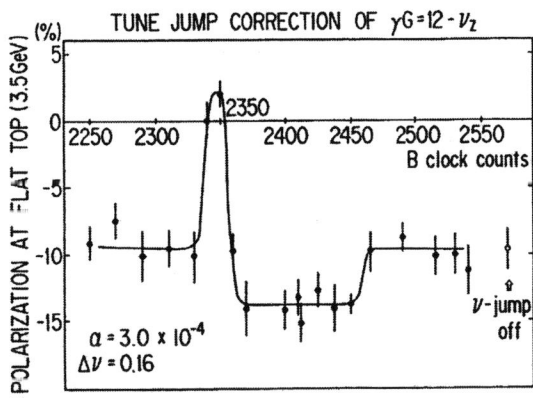

FIGURE 4. Correction curve of a strong intrinsic resonance, $\gamma G = 12 - v_z$.

SLOW PASSAGE FOR STRONG INTRINSIC RESONANCES

Although it was expected that the resonances, $\gamma G = \nu_z$ and $4+\nu_z$, were strong as described before, some depolarization was observed. It was planned to cross these resonances with the slow passage method by adjusting the falling tail of the pulsed quadrupole magnets' field to fit the gradient of γG. As shown in Fig. 5, the beam polarization was improved by this technique.

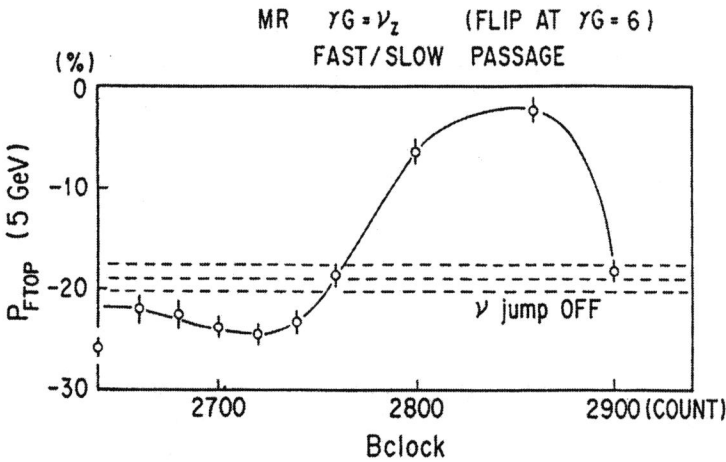

FIGURE 5. Correction curve of the strong intrinsic resonance, $\gamma G = \nu_z$.

MULTIPLE RESONANCE CROSSING BY SYNCHROTRON OSCILLATION

In the booster, the intrinsic resonance, $\gamma G = \nu_z$, was very strong and an adiabatic spin-flip was expected without significant depolarization. However, it was found to cause large depolarization because of the betatron and spin tune modulations caused by the synchrotron motion. In the booster, the synchrotron oscillation was several kHz, which was fast enough to cause multiple crossings, although the RF voltage for the polarized beam was lower than the normal RF level for the high intensity (unpolarized) operation. This effect was observed as the RF voltage dependence of the polarization, as shown in Fig. 6. In the first acceleration test, almost complete depolarization was observed for a typical RF voltage level (20 kV peak).

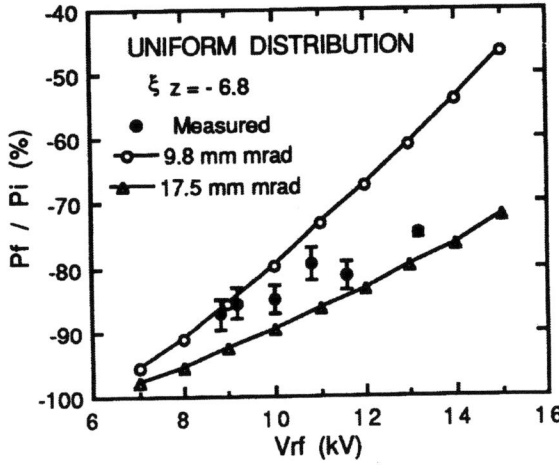

FIGURE 6. RF voltage dependence of the beam polarization at 500 MeV.

Figure 7 shows a simple view of resonance crossing with and without synchrotron oscillation [13]. As shown in Fig. 7-1, the spin tune without the synchrotron oscillation is linearly increasing, $\gamma G = \gamma_0 G + \alpha\theta$, which crosses the resonance line, $\gamma G = nN - \nu_{z0}$ (=constant). Here, α and θ are the crossing speed and azimuthal variable, respectively. Taking the synchrotron oscillation into account, the equations become:

$$\gamma G = \gamma_0 G + \alpha\theta + \gamma_0 G \beta^2 (\Delta p/p) \cos(\nu_s\theta+\delta), \quad (3)$$
$$nN - \nu_z = nN - \nu_{z0} \pm \xi(\Delta p/p) \cos(\nu_s\theta+\delta), \quad (4)$$

where ν_s and δ are the synchrotron tune and phase of oscillation, respectively. In the booster, ν_s and α were about 0.01 and $1.9 \cdot 10^{-6}$ at the intrinsic resonance, $\gamma G = \nu_z$. The equations give

$$\gamma G - (nN - \nu_z) = \alpha\theta + (\gamma_0 G \beta^2 \pm \xi)(\Delta p/p) \cos(\nu_s\theta+\delta). \quad (5)$$

Figure 7-2 shows the schematic of the multiple crossing. Equation (5) then suggests that the chromaticity correction should satisfy,

$$\xi = \pm \gamma_0 G \beta^2. \quad (6)$$

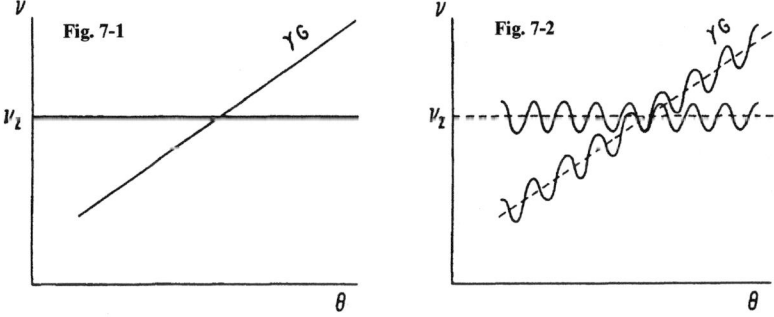

FIGURE 7. Resonance crossing without (left) and with (right) the synchrotron oscillation.

Two pulsed sextupole magnets were installed to correct the chromaticity during the resonance crossing. The results are shown in Fig. 8. The polarization at 500 MeV had the maximum value at the sextupole current of 50 A. The polarization survival was improved from 75% to 82 % by the correction.

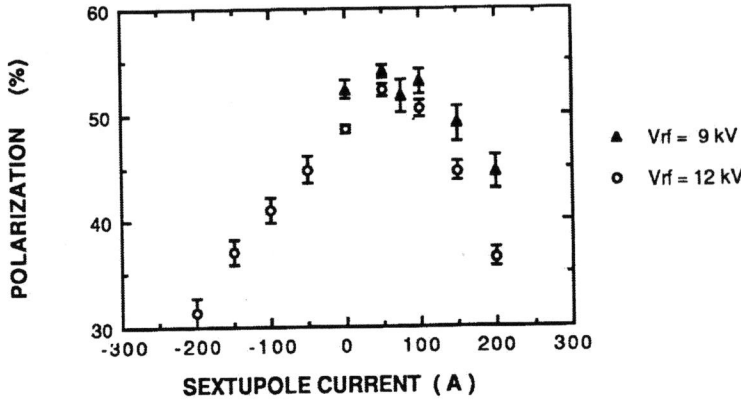

FIGURE 8. Sextupole current dependence of the $\gamma G = \nu_z$ resonance.

CONCLUSIONS

Polarized proton beam had been accelerated at the KEK-PS. Some depolarization resonances were passed using both tune jump and the spin-flip techniques. The effects caused by the synchrotron side band had been observed at the strong intrinsic resonances.

After the polarized deuteron acceleration in 1996 [14], the polarized project at the KEK-PS was terminated. KEK then decided to dedicate its efforts for the long baseline neutrino oscillation experiment (K2K) and all the equipment for the polarized beam acceleration program was evacuated in order to install instruments required to obtain a high intensity beam.

REFERENCES

1. Y. Mori et al., in *Proc. of the 5th Int'l. Symp. on High Energy Spin Physics*, BNL (1982), p. 611.
2. S. Hiramatsu et al., in *Proc. of the 6th Int'l. Symp. on High Energy Spin Physics, Marseilles* (1984), p.C2_529.
3. S. Hiramatsu et al., in *Proc. of the 7th Int'l. Symp. on High Energy Spin Physics*, Protvino (1986), p. 119.
4. T. Toyama et al., "Chromaticity Correction for an Intrinsic Depolarization Resonance", in *Proc. of the 14th Int'l. Conf. on High Energy Accelerators*, Tsukuba, (1989), p.167.
5. H. Sato et al., in *Proc. of the 6th Symp. on Acc. Sci. and Tech.* (1987), p.189
6. H. Sato et al, "Acceleration of the Polarized Proton Beam at the KEK 12 GeV PS", *Nucl. Instrum. Methods* **A272**, 617 (1988).
7. S. Fukumoto et al., KEK-Preprint 86-49 (1986).
8. H. Sato et al., "Depolarizing Resonance Correction in the Polarized Proton Beam Acceleration up to 5.0 GeV at the KEK PS", *Jpn. J. Appl. Phys.* **27**, 1022 (1984).
9. Igarashi et al., in *Proc. of the 5th Symp. on Acc. Sci. and Tech.* (1984), p. 134.
10. C. Ohmori et al., "Proton Polarimeter for High Energy Experiments at KEK", *Nucl. Instrum. Methods* **A278**, 705 (1989).
11. H. Sato et al., "Internal Polarimeter for the Polarized Proton Beam at the KEK 12GeV PS", in *Proc. of the 8th Int'l. Symp. on High Energy Spin Physics,* 1988, AIP Conf. Proc. **187**, 1355 (1989).
12. H. Sato et al., "Pulsed Quadrupole Magnet System for the Polarized Beam Acceleration at the KEK 12 GeV PS", KEK Preprint 87-22 (1987).
13. S. Hiramatsu et al., "Depolarizing Problem at the KEK 12GeV PS", in *Proc. of the 8th Int'l. Symp. on High Energy Spin Physics,* Minneapolis (1988), p. 1436.
14. H. Sato et al., "Polarized Deuteron Beam Acceleration at the KEK-PS", *Nucl. Instrum. Methods* **A385**, 391 (1997).

AGS Pulsed Quadrupoles: History and Future[1]

A. D. Krisch

Spin Physics Center, University of Michigan, Ann Arbor, MI 48109-1120 USA

Abstract. This paper contains a brief discussion of the AGS polarized beam work in the 1980's. It also suggests that it might be wise to again use the old pulsed quadrupoles to overcome the problematic weak intrinsic depolarizing resonances.

Leif Ahrens earlier presented a very nice talk on an analysis of the results from the AGS polarized proton beam runs of the 1980s. I would like to now discuss some of the hardware that was used in these runs, with the goal of seeing if any of it could be used to help increase the AGS polarization in this Millennium. Using some existing hardware has many advantages, especially when a run is approaching. A discussion of this AGS polarized beam hardware can be found in great detail in a long Physical Review paper [1], published in 1989, and based on F. Z. Khiari's PhD Thesis. I will not reproduce here the many figures of the hardware or the data from their use, since they can be found in this paper.

Instead I will focus on the possible future use of one hardware item: some of the twelve pulsed quadrupoles. These 1.6-μs-rise time ferrite quadrupoles and their 20 MW power supplies were certainly the most difficult and expensive part of the $10 Million AGS Polarized Proton Beam Project. Moreover, they did not do a completely adequate job of overcoming the strong intrinsic depolarizing resonances at the AGS; however, they were rather successful in overcoming the weak intrinsic polarizing resonances [1]. I stress this point because the new rf dipole technique, developed in the past few years by this Millennium's AGS polarized beam team [2], has been fairly successful at overcoming the AGS's strong intrinsic depolarizing resonances, but less successful at overcoming its weak intrinsic depolarizing resonances. Thus, perhaps one should now consider:

- using the old pulsed quadrupoles to overcome the weak intrinsic depolarizing resonances;
- using the new rf dipole technique only for the strong intrinsic depolarizing resonances, where it seems to work rather well.

The new technique uses an rf dipole, whose frequency is moved very close to a strong intrinsic depolarizing resonance. This proximity enhances the strength of the

---

[1] Supported by a Research Grant from the U.S. Department of Energy

strong intrinsic depolarizing resonance enough to allow it to spin-flip the beam almost totally, thus maintaining almost all the beam polarization. If the intrinsic depolarizing resonance is fairly strong, then one does not have to move the rf frequency too close to it; thus the beam is not destroyed. However, to use this technique for a weak intrinsic depolarizing resonance, one must move the rf frequency very close to the weak intrinsic depolarizing resonance; then, it becomes difficult to avoid a beam blow-up resonance, which can destroy all or part of the beam. Thus for the weak intrinsic depolarizing resonances, one is caught between two competing goals:
- maintaining the beam polarization;
- maintaining the beam intensity.

Even with perfect AGS betatron tune stability, this technique may be difficult for routine operation.

Since this new technique's problem is most serious for the weak intrinsic depolarizing resonances, and since the still-existing 1.6-μs-rise time ferrite quadrupoles worked rather well for them [1], it seems wise to consider using these quadrupoles for this part of the job. I think that I first discussed this idea with Mei Bai and Andreas Lehrach during the SPIN 2002 Excursion to the Museum of Natural History in Manhattan; this possibly valuable discussion might be mentioned the next time anyone objects to excursions at Symposia.

The main problem with using the pulsed quadrupoles, is that their 22 MW power supplies are in even worse shape than they were in the 1980s, when they caused many problems for many people, especially Larry Ratner and me. Fortunately, one can overcome the weak intrinsic depolarizing resonances with a fairly small betatron tune jump and thus a fairly slow rise time. Therefore, some of the old 22 MW power supplies might work well enough to operate at a few MW; or, if they are completely dead, it might not cost too much to replace them with some much lower power devices.

Therefore, I prepared Table 1, which demonstrates several different possible plans for using three different techniques for overcoming the three main polarization problems [3] in the AGS:
1. many imperfection depolarizing resonances;
2. some weak intrinsic depolarizing resonances;
3. some strong intrinsic depolarizing resonances.

In a way it is sad that one cannot find a way to find a single elegant device, such as the Siberian snake, to overcome all depolarization problems at the AGS. However, even Yaroslav Derbenev, who is attending this Workshop cannot always produce miracles on demand. Unfortunately, the AGS is at just the wrong energy for a full Siberian snake:
- its injection energy is too low for a practical helical or dipole snake to fit in its straight sections;
- its maximum energy is too high for a solenoid snake to fit in its straight sections.

Thus, in the absence of any new miracle, one should seriously consider, the less elegant, but hopefully practical solution of using *three techniques* for *three problems*.

TABLE 1. Overcoming the AGS Depolarizing Resonances

HISTORY

	~ 40 IMPERFECTION	~ 6 INTRINSIC
1980s	96 Correction Dipoles	12 Pulsed Quads
1990s-2002	5% Warm Solenoid Snake	rf Dipole

2003-2004

	~40 IMPERFECTION	~3 STRONG INTR.	~3 WEAK INTR.
Plan 1	20-25% Cold Helical Snake	rf Dipole?	
Plan 2	5% Warm Solenoid Snake	rf Dipole	4-8 Pulsed Quads
Plan 3	5% Warm Helical Snake	rf Dipole	4-8 Pulsed Quads
Plan 4	5% Ramped Warm Solenoid Snake	rf Dipole	4-8 Pulsed Quads

REFERENCES

1. Khiari, F.Z. *et al.*, *Phys. Rev.* **D39**, 45-85 (1989).
2. Roser, T. *et al.*, First talk in this Workshop and references therein.
3. Note added in proof: There may also be some coupling resonances, which Thomas Roser considered in choosing the schemes described in the Summary Talk.

OPPIS Upgrade for 2003 Polarized Run in RHIC[a]

A. Zelenski[*], J. Alessi[*], B. Briscoe[*], A. Kponou[*], S. Kokhanovski[†], V. Klenov[†], V. LoDestro[*], J. Ritter[*], V. Zubets[†]

*Brookhaven National Laboratory, USA
†INR, Moscow, Russia

Abstract. The polarization dilution by molecular ions which are produced in the ECR primary proton source is discussed. The molecular component can be reduced to about 5% by ECR source-operation optimization. It is further suppressed by optimization of the extraction electrode optics and by the decelerating einzel lens in the 35 keV LEBT line. As a result, the proton polarization of the accelerated beam was increased to over 80%, as measured in the 200 MeV proton-deuterium polarimeter. The OPPIS upgrade for 6 2/3 Hz repetition rate operation is also discussed.

INTRODUCTION

The RHIC OPPIS produces routinely 0.5-1.0 mA (maximum 1.5 mA) current in a 400 μs pulse duration. Polarized H⁻ ions are produced in the OPPIS at 35 keV beam energy. The beam is accelerated to 200 MeV with an RFQ and a linac for charge exchange strip-injection into the Booster. About 50% of the OPPIS beam intensity can be accelerated to 200 MeV. The 400 μs H⁻ ion pulse is captured in a single Booster bunch which contains about $4 \cdot 10^{11}$ polarized protons. The single bunch is accelerated in the Booster to 1.5 GeV kinetic energy and then transferred to the AGS, where it is accelerated to 25 GeV for injection to RHIC.

The OPPIS' initial longitudinal polarization is converted to the transverse direction while the beam passes two bending magnets. The second 47.4° bending magnet switches linac injection between polarized and unpolarized high intensity (up to 100 mA) H⁻ ion beam. The magnet is pulsed and either beam can be accelerated pulse-to-pulse in the same RFQ. A pulsed focusing solenoid in front of the RFQ is tuned for optimal transmission for either beam. It rotates the polarization direction for about 420°, but still keeps it in the transverse plane, and a final polarization alignment to the vertical direction can be adjusted by a spin-rotator solenoid in the 750 keV beam transport line before injection to the linac [2]. The AGS cycle for polarized beam operation is 3 seconds. The OPPIS operates at 1 Hz repetition rate and additional source pulses were directed to the 200 MeV p-Carbon polarimeter for continuous polarization monitoring by switching of another pulsed bending magnet in the high-energy beam transport line. The spin-rotator tuning is done using vertical polarimeter arms.

OPPIS PERFORMANCE

Recent source component upgrades have significantly improved the OPPIS performance and reliability. The OPPIS polarization technique is described elsewhere [3,4]. The schematic OPPIS layout is presented in Fig.1.

The ECR-primary proton source upgrade: The ECR operation in a pulsed mode was studied at 1-7 Hz repetition rate and 5-100 ms pulse duration. A significant reduction in the optimal hydrogen feeding flow was observed, which might be explained by gas adsorption on the quartz tube, then desorption at the beginning of the RF pulse. This produces additional gas contribution sufficient to maintain an optimal density for about 5-10 msec at the beginning of the pulse. Pulsed ECR operation reduces gas load to the vacuum system and heat load to the ECR and Rb cell. In dc operation an oxygen gas admixture is required to optimize the ECR-source proton production and proton to molecular H_2^+ ions dissociation ratio. In the pulsed operation sufficient amount of oxygen is supplied to the discharge from residual gas.

Advantages of the pulsed operation were not fully used in the RHIC run, because the heat load difference between pulsed and dc operation eventually caused leaks in a quartz to copper cavity seal. In the new ECR cavity silicone O-rings (silicone has lower RF-power absorption than other rubber –like materials) are used instead of a teflon-indium seal. The silicone O-ring exposure to 29 GHz microwave power is reduced by the seal design. The quartz tube air-cooling was improved to prevent seal overheating through the contact with the hot quartz tube.

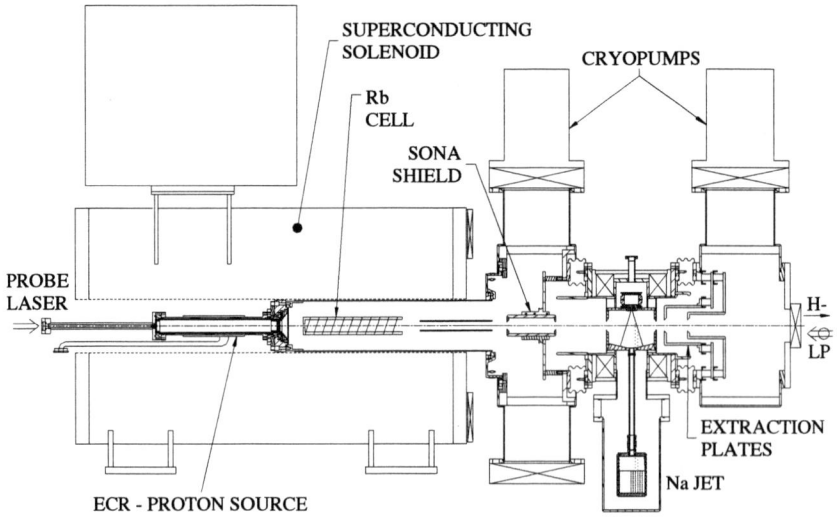

FIGURE 1.: The RHIC OPPIS layout.

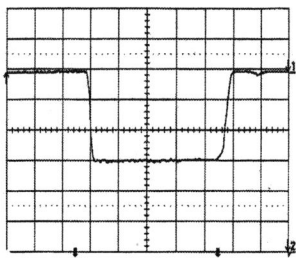

FIGURE 2. The polarized H- ion beam current pulse in a 35 keV LEBT Faradey cup.
Vertical scale-500 µA/div, horizontal-100µsec/div.

The silicone O-rings provide the flexibility to compensate the difference in thermal expansion between quartz and copper, which allows frequent switching between dc and pulsed modes for optimal source operation. The new ECR-cavity length was also increased to match the magnetic field shape in the ECR-discharge region. As a result the maximum polarized H⁻ ion current was increased to 1.5 mA (see Fig. 2).

Laser system upgrade: The laser pulse duration was extended to 450 µs for optical pumping during 400 µs H⁻ current pulse. On-line rubidium polarization measurements were implemented, by probe laser linear polarization rotation measurements (Faraday rotation polarimeter). These measurements give a reliable polarization readout for confirmation of the spin sequence pattern, which is injected to RHIC. Recently, the laser was successfully tested at a 7 Hz OPPIS repetition rate.

POLARIZATION STUDIES

Over 85% polarization was obtained during the final RHIC OPPIS tests at TRIUMF where only electrostatic beam optics were used [4]. A 70-72% polarization was measured in the first year 2000 run at 200 MeV in the p-carbon polarimeter, at a reduced intensity of less than 10 µA. (The polarimeter was designed for the low current atomic beam source and detectors where greatly overloaded at 200 µA current.) The polarimeter was upgraded for high current operation by extending target to detector distance from 70 to 250 cm and additional collimator installation to suppress particles scattered from the target holder. Polarization measurements at 180 µA beam intensity were obtained during spin-rotator tuning. Still there is some saturation, and measured polarization is higher at 120 µA. The p-carbon polarimeter is inclusive and relied on the calibration measurements, which were done 10 years ago under different conditions. The absolute polarization is an important reference point for polarization loss measurements in Booster and AGS and for evaluation of the OPPIS development status. An upgrade was suggested an upgrade to a polarimeter using p-deutron elastic scattering, where the analyzing power is precisely known at 200 MeV/5/. Four additional detector arms for the proton-deutron coincidence measurements from CD_2

target (deuteriated polyethylene) were installed in the summer of 2001. During the November 2001-January 2002 run, p-C and p-D scattering data were accumulated and the old analyzing power for 12°degree inclusive pC scattering of a 0.62 was closely reproduced [6]. Therefore the 70-75% polarization at 200 MeV was real and lower than expected.

POLARIZATION DILUTION BY H_2^+ MOLECULAR IONS

There exists a molecular H_2^+ ion component in any plasma ion source. In the OPPIS, molecular ions after dissociation will appear as H⁻ ions with the half of the primary beam energy. The polarization of this beam might be different from the main beam (measurement in Lamb-shift polarimeter gives about half the polarization for this molecular component). This component was observed at the TRIUMF OPPIS, but it was efficiently suppressed by electrostatic lenses in the 3 keV LEBT. In the RHIC OPPIS the H⁻ beam is accelerated for 32 keV immediately after ionization, producing 35 keV main beam and 33.5 keV beam from molecular ion admixture. These beams are not well separated in the LEBT, and the molecular component is responsible for polarization dilution. At lower acceleration energy these beams are separated, and the half energy component was directly observed (see Fig.3).

A value of molecular component of about 10-40 % was measured under different ECR conditions. Since every H_2^+ ion is dissociated to two half energy atoms, it means 5-20% molecular component out of the ECR-source. The molecular component increases at higher ECR extraction voltage. At 4.0 keV its value is about 40%. The H⁻ yield drops at atomic beam energy above 3.0 keV, but for half energy beam of a 1.5 – 2.0 keV the yield is at maximum value. It explains an increase of molecular component up to 40% at 4.0 keV ECR extraction energy, and correspondent polarization decrease, which was also observed in polarization measurements.

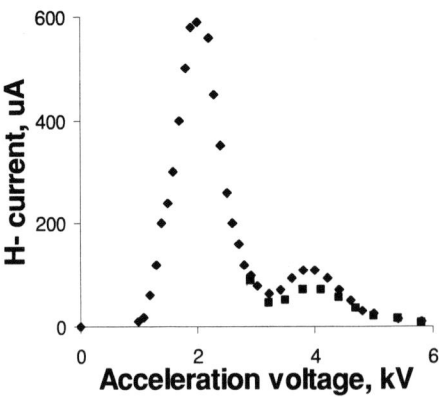

FIGURE 3. Molecular H_2^+ beam component is appeared as a second bump shifted to about a half primary ECR proton energy at a fixed bending magnet setting. Diamonds-dc operation; squares-pulsed.

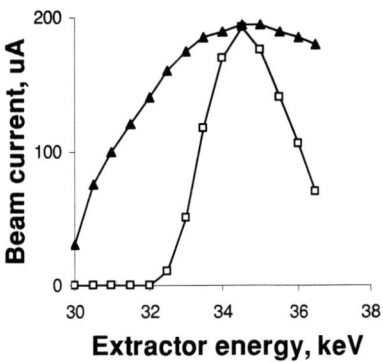

FIGURE 4. Linac transmission vs extractor voltage (acceleration voltage applied to the jet-ionizer cell). Triangles-LEBT optics with the magnetic quadrupole triplet; squares-quad triplets replaced with the decelerating Einzel-lens.

The increase of the ECR current was obtained earlier with an admixture of a few percent of gaseous oxygen to the hydrogen in the source [2]. The oxygen admixture also reduces the molecular H_2^+ ion production i.e. improves the dissociation ratio in the source. The magnetic field shape in the ECR region also affects the dissociation ratio. The optimization of the ECR source parameters gave rise to an increase of the main beam intensity and reduction of the half energy, lower polarization component to below 10%. The ECR operation in a pulsed mode was also studied. A significant molecular component suppression was also observed in a pulsed operation (see Fig.3), due to a difference in the rise time for main and half energy beam components. In a pulsed operation the molecular component is about 5%. As a result the polarization at 200 MeV was increased to 75%. The next step was a suppression of the lower energy component during acceleration to 35 keV after the ionizer and in the LEBT. The two-gap acceleration system was upgraded to three-gaps and the voltage in the first gap was tuned to suppress the half energy component. The voltages at the second and third gaps were adjusted to minimize the main component losses. The triplet of magnetic quadrupole lenses at the OPPIS exit was replaced by a single deceleration einzel lens.

The combined effect of these modifications is a significant (almost ten times) suppression of the beam transmission in LEBT for the lower energy molecular origin beam (see Fig.4). As a result of these upgrades, a polarization at 200 MeV of 80-82% was measured in both p-Carbon and p-Deutron polarimeters.

OPPIS UPGRADE FOR 6 2/3 Hz REPETITION RATE

A multi-bunch beam is required for depolarization studies during the AGS energy ramp with a new proton-carbon CNI polarimeter [5]. A four OPPIS pulses will be injected to AGS at 6 2/3 Hz repetition rate. In 2000 and 2001-02 polarized runs the

OPPIS was operated at 1 Hz repetition rate and additional pulses between injection to AGS were directed to 200 MeV polarimeter. A new mode of operation requires a number of upgrades in the source, polarimeter, control and data acquisition system.

The ECR is operated either dc or pulsed, so the higher rate is not a problem. The pulsed high-voltage bias (32.0 kV) of the sodium-jet ionizer cell was a major concern. The ionizer operation at 1 Hz was reliable, but a quite high (up to 400 mA) magnetron-type discharge in the crossed electrical and magnetic fields was observed during the high-voltage pulse. This caused some outgasing, and at higher rate might increase the probability of sparking. An insufficient superconducting solenoid field confinement by the yoke was identified as a problem. A new magnetic field correction coil was attached to the solenoid yoke (see Fig.1), which suppressed almost completely the stray magnetic field in the ionizer high-voltage gap. As a result, the discharge current was reduced at least 10 times and a stable spark-free operation was obtained at 6 2/3 Hz. At the same time the Sona-transition conditions were also greatly improved due to magnetic field gradient reduction in the zero-crossing region.

The laser cooling system was improved, and stable operation at 6 2/3 Hz was obtained. While four pulses will be produced in less than 0.5 s, which is followed by about 3 second AGS ramp time, the laser must be pulsed continuously to maintain stable laser power and wavelength operation. The laser service periodicity and lifetime at high rate still have to be studied. For laser circular polarization reversal at 1 Hz repetition rate, $\lambda/4$ were plates were inserted in the laser beam by an air-driven slide. Due to large Zeeman shift, the optical pumping can be done by the linear polarized light, which is a combination of both circular polarization components (the wrong component is "sterile" and does not produce any transition, but effective laser power is reduced in half). About 2-3 % polarization losses were obtained experimentally without the $\lambda/4$ plates at optimal laser tune conditions. It is planned to install a Pockels-cell for laser polarization reversal at high rate.

A Faraday-rotation polarimeter for spin-sequence confirmation now operates routinely at 1 Hz and will be used for the coming run. A DAQ upgrade for higher repetition rate is in progress for polarization measurements in the source and in a 200 MeV polarimeter.

REFERENCES

a. Work supported by the U.S. Department of Energy
1. G. Bunce et al., "Polarized protons at RHIC", *Particle World* **3**, 1 (1992).
2. A. Zelenski et al., *Rev. Sci. Instrum.* **73**, 888 (2002).
3. A. Zelenski et al., *Nucl. Instrum. Methods* **A402**, 185 (1998).
4. A. Zelenski, in *Proc. of SPIN 2000, AIP Conf. Proc.* **570**, 179 (2001).
5. E. Stephenson, private communications.
6. H. Huang et al., to be published in *Proceedings of EPAC 2002*.

AGS Lattice Changes to Eliminate Weak Intrinsic Resonances

A. Lehrach* and V. H. Ranjbar[†]

Forschungszentrum Jülich GmbH, 52425 Jülich, Germany
[†]*Fermi National Accelerator Laboratory, Batavia, IL 60510, USA*

Abstract. Significant polarization losses are observed at weak intrinsic spin resonances in the AGS. The suppression of weak intrinsic resonances is explored using a family of 12 quadrupoles in the AGS. Experimental results using the existing family of 12 horizontal and vertical tune quadrupoles are presented and theoretical results for a possible extra family of 12 quadrupoles in the 15th lattice straight section of each superperiod are discussed.

INTRODUCTION

The goal of the Relativistic Heavy Ion Collider (RHIC) spin project at Brookhaven National Laboratory is the delivery of polarized protons at energies of 250 GeV. A major bottle neck in this process is the Alternating Gradient Synchrotron (AGS) which accounts for most of the polarization losses during the chain of acceleration leading to RHIC.

Spin depolarizing resonances in the AGS are of three main types, imperfection resonances, intrinsic resonances and coupled spin resonances. All three are a result of spin kicks accumulated from vertical beam motion through quadrupoles. They differ in the causes of the vertical motion. Imperfection resonances arise from vertical closed orbit distortions. Intrinsic resonances arise from vertical betatron oscillations. Coupled spin resonances are derived from the projection of horizontal betatron oscillations in the vertical plane due to linear coupling. Since imperfection resonances accumulate from uncorrelated periodic vertical orbit kicks, the resonance condition occurs when the spin tune v_{sp} becomes an integer value. Since intrinsic resonances are due to vertical betatron oscillations, this resonance condition occurs when the spin tune becomes a harmonic of the vertical betatron tune v_z, i.e. $v_{sp} = N \pm v_z$ (where N is an integer). Likewise since coupled spin resonances arise from the horizontal betatron tune v_x, these resonances occur whenever $v_{sp} = N \pm v_x$.

In recent years several novel methods to overcome spin depolarizing resonances in the AGS have been pioneered. In 1994 a partial snake was installed and tested in the AGS [1]. The partial snake was designed to rotate the spin vector by several degrees each turn. The result is the creation of a strong resonance at every $G\gamma = integer$, where $G = \frac{g}{2} - 1$ is the anomalous magnetic moment coefficient and γ the Lorentz factor. Since the resonance induced is strong enough to flip the spin vector completely, polarization can be preserved across the imperfection resonances. However, since the partial snake was built using a solenoid, strong coupling was introduced into the AGS and the coupled spin resonances were enhanced. In 1999, an ac dipole was installed and used to overcome strong intrinsic resonances in the AGS [2]. By driving the ac dipole near the vertical betatron tune and thus enhancing the natural intrinsic resonance, a full spin flip is achieved. However, the ac dipole could not be used on the weak intrinsic resonances since the amplitude of the beam oscillation required to achieve a full spin flip was beyond the physical aperture of the AGS beam pipe.

To date, problems in the AGS remain with the weak intrinsic resonances and the coupled spin resonances. Problems with these resonances have hampered efforts to deliver the required 70% polarization for RHIC. The AGS has only been able to deliver a maximum of 40% polarization to RHIC. The polarized proton source during the 2002 run, delivered protons polarized at 70% to the AGS booster at 2×10^{11} particles per shot. Without polarization loss, the polarized proton bunches was transferred to the AGS at $G\gamma = 4.6$. Thus losses in the AGS amount to nearly 40% of the delivered polarization, all of which are due to the coupled and weak intrinsic resonances. In this paper we explore the response of the weak intrinsic resonances to several families of quadrupoles. In the first section we begin by taking a close look at the inherent intrinsic resonances structure of the AGS lattice and consider the effect of an additional family of 12 quadrupoles. In the second section experimental results using the family of horizontal tune quadrupoles are presented. Finally in the last section we present theoretical results using an additional family of quadrupoles in the 15th straight section of each superperiod.

THE AGS INTRINSIC RESONANCE STRUCTURE

It is well know that the spin resonance strength can be approximated by [3]

$$\varepsilon_K \approx -\frac{1+G\gamma}{2\pi} \oint z'' e^{iK\theta} ds = +\frac{1+G\gamma}{2\pi} \oint k_z(s) z e^{iK\theta} ds. \qquad (1)$$

Here we have used the homogeneous equation $z'' = -k_z(s)z$ with $k_z(s)$ given as the focusing function. For intrinsic resonances we can use the equation for betatron motion and Eq. (1) becomes

$$\varepsilon_K \approx \frac{1+G\gamma}{2\pi} \sqrt{\frac{\varepsilon_N}{\pi\gamma}} \oint k_z(s) \sqrt{\beta_z(s)} \cos(v_z \phi_z(s)) e^{iK\theta} ds. \qquad (2)$$

Here β is the betatron function, ε_N the normalized emittance, ϕ the betatron phase and θ the angular location in the ring.

The AGS lattice is made up of twelve superperiods each containing twenty combined function magnets of long and short lengths. In Fig. 1 a graphical representation of the lattice is shown. The resulting structure proves fairly complex. But some general observations are possible. The lattice can be broken down into two sections which are anti-symmetric. Further these sections can be broken down into a total of four sections of two anti-symmetric pairs. There are also two mirror symmetric pairs. While clearly the overriding periodicity of 12 places all our intrinsic resonances at $12n \pm v_z$, the anti-symmetric structure can explain the odd and even substructure of the resonances. This

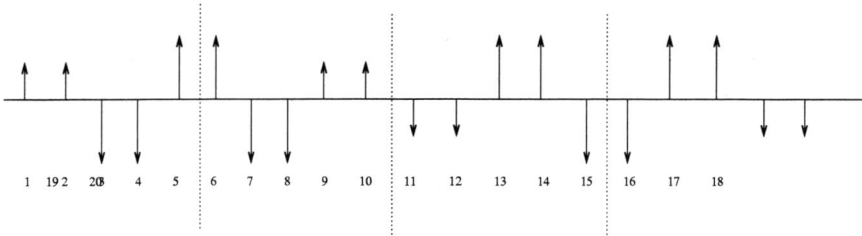

FIGURE 1. AGS superperiod. The AGS is composed of 12 superperiods, each consists of 20 combined function dipoles. Here the up and down vectors show the direction and magnitude of the focusing gradients. Two tuning quadrupoles are located at position 3 and 17.

explanation follows in a manner similar to the evaluation of a typical FODO lattice structure. In a standard FODO lattice the Focusing and Defocusing elements (which are anti-symmetric pairs) can contribute additively to the spin kick when the resonance is odd ($n =$ odd) [3]. Applying this logic to the AGS lattice we can see that the middle two anti-symmetric pairs can be viewed as having an overall sign difference when seen as a whole. The same is true for the outer two anti-symmetric pairs. For the AGS we know that $\phi \approx \theta$ so between the anti-symmetric pairs there will always be phase difference of $(K \pm v_z)\frac{m\pi}{12}$. Here m is some integer determined by which pair is being counted. So at the intrinsic resonance where $K = 12n \pm v_z$, the anti-symmetric elements should contribute additively to the spin kick when n is odd and subtract when n is even. This explains why we see in Table 1 that the odd resonances tend to be larger than the even ones (except when $n = 0$).

TABLE 1. Resonance strengths ε_k as calculated by DEPOL [4] for the bare AGS machine using the rms emittance for a 10 π mm-mrad beam.

γ	$G\gamma$	Re(ε_k)	Im(ε_k)	$|\varepsilon_k|$
4.871	0+v	0.006128	-0.000271	0.00613421
8.515	24-v	0.000240	0.000011	0.00023988
11.565	12+v	0.000094	0.002118	0.00211998
15.208	36-v	-0.000248	0.005589	0.00559465
18.258	24+v	0.000507	-0.000022	0.00050779
21.902	48-v	0.000652	0.000029	0.00065232
24.951	36+v	0.000479	0.010820	0.01083078

Suppression of the Weak Intrinsic Resonances

The complexity of the lattice seems to prohibit a real increase in the superperiodicity of the lattice without significant and costly re-modifications. While there exist many points of overall mirror symmetry, it is clear that neither a single nor several quadrupoles can increase the overall periodicity. However, it is well known that in certain instances using carefully placed families of quadrupoles that individual resonances can be suppressed by the introduction of a countervailing perturbation to the resonance strength which can be approximated by [3],

$$\Delta \varepsilon_K = (\frac{1+G\gamma}{4\pi})\sqrt{\frac{\varepsilon_N}{\pi\gamma}} \int \Delta g(s) \sqrt{\beta} (e^{i(\nu\phi-K\theta)} + e^{-i(\nu\phi-K\theta)}) ds. \qquad (3)$$

We use $\Delta g(s) = \frac{1}{B\rho}\frac{\partial B_z}{\partial x}$ as the focusing strength. This equation will give a good approximation providing that both $\Delta g(s)$ and $\Delta \beta(s)$ are small. It seems that if we use just the right field strengths and locations, we can use this perturbation to cancel our existing resonance. This is known as spin matching. This method of spin matching or harmonic suppression has been proposed for other machines [5, 6] and actually successfully implemented at COSY [7].

It might prove insightful if we try to develop an analytical approximation to Eq. (3). If we consider the effect of only one additional quadrupole per superperiod, Eq. (3) can be integrated as a sum over the number of superperiods in the lattice. Using a thin lens approximation this series can then be summed using the properties of a geometric series to give,

$$\Delta \varepsilon_K = (\frac{1+G\gamma}{4\pi})\sqrt{\frac{\varepsilon_N}{\pi\gamma}} \{ e^{i(\nu_z+K)(\frac{P-1}{P})\pi} \zeta_P(\frac{K+\nu_z}{P})[g_1\sqrt{\beta_1}e^{i(\nu_z\phi_1+K\theta_1)}] + $$
$$e^{i(K-\nu_z)(\frac{P-1}{P})\pi} \zeta_P(\frac{K-\nu_z}{P})[g_1\sqrt{\beta_1}e^{i(K\theta_1-\nu_z\phi_1)}] \}. \qquad (4)$$

Here θ_1 and ϕ_1 are the angle and betatron phase at the quadrupole's position. We also assume that we are inserting one quadrupole per superperiod (P). Additionally, we have made use of $\zeta_N(x)$ [3] which is defined as

$$\zeta_P(x) = \frac{\sin P\pi x}{\sin \pi x}. \qquad (5)$$

For $x = N$ ($N = 0, \pm 1, \pm 2, ...$), $\zeta_P(x)$ goes to $(-1)^N P$ for even P and P for odd P. In the AGS where $\theta \approx \phi$ we can simplify Eq. (4) for the intrinsic resonance condition ($12n \pm \nu_z$) to obtain

$$\varepsilon_K \approx (\frac{1+G\gamma}{4\pi})\sqrt{\frac{\varepsilon_N}{\pi\gamma}} 12(-1)^n g_1 \sqrt{\beta_1} e^{in(11)\pi+12n\phi}. \qquad (6)$$

From this expression it appears that it should be possible to use vertical and horizontal tune quadrupoles to effect the weak spin resonances. These quadrupoles are located at the 3rd ($\phi = 4.5°$) and 17th ($\phi = 25.5°$) straight section of each superperiod. Since for the weak intrinsic resonances, $K = 24 \pm \nu_z$ and $48 - \nu_z$ (n=2,3) both yield imaginary and real resonance contributions in Eq. (4) it would be necessary for both the horizontal and vertical quadrupoles to be activated at once. In this way a configuration could be constructed where either the imaginary or real part of the perturbation would cancel out leaving only a single real or imaginary component. However calculations have shown that the field strength requirements to achieve this would require leaving the 8.5-9.0 vertical and horizontal tune space and further require a field strength beyond the capacity of the existing quadrupoles.

COMPARISON WITH EXPERIMENTAL RESULTS

However, within the existing tune box of 8.5-9.0 there should exist some optimal tune setting which would minimize the effect of the weak intrinsic resonances. To help determine this setting, during the 2002 polarized proton run we had the opportunity to gather data on the response of the weak intrinsic spin resonances in the AGS to various tune settings. What follows is a presentation of this data compared with calculated results using the enhanced version of DEPOL [8] in conjunction with MAD.

In order to discern the effects of each individual resonances we proceeded by adjusting the AGS main magnet function to ramp up to three successively higher flattop values, $G\gamma =18.5$, 34.5 and 41.5. In Figs. 2 to 4 we plot the response for each of the weak intrinsic resonances to changes horizontal tune along with DEPOL calculated values. Since we did not have the use of a horizontal tune meter throughout the whole 2002 run, all DEPOL calculations were made using the measured input currents to the horizontal and vertical tune quadrupoles. The tunes quoted in all the AGS graphs are set point tunes. Since much work in the past has been devoted to developing an accurate model of the AGS lattice for use in MAD, tune calculation performed in the MAD model present a good representation of the actual tunes in the AGS.

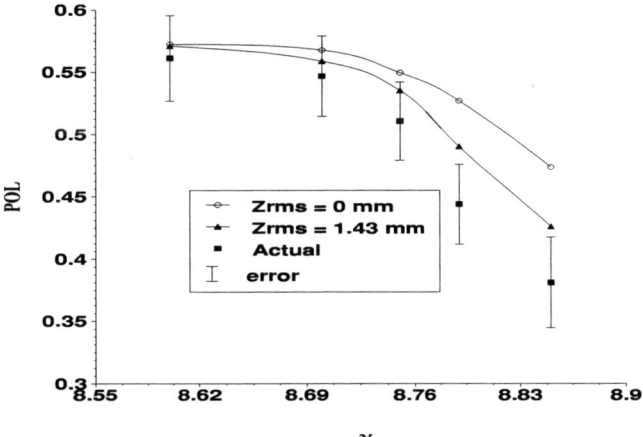

FIGURE 2. $24 - \nu$ resonance crossing at fixed 8.8 vertical tune, as a function of horizontal tune measured at $G\gamma = 18.5$ flattop plotted with DEPOL predictions assuming initial polarization was the same as measured at $G\gamma = 12.5$, 58%. Plots are with and without added closed orbit distortions $Z_{rms} = 1.43$mm and $X_{rms} = 0.067$mm. Emittances in vertical were 10π mm-mrad.

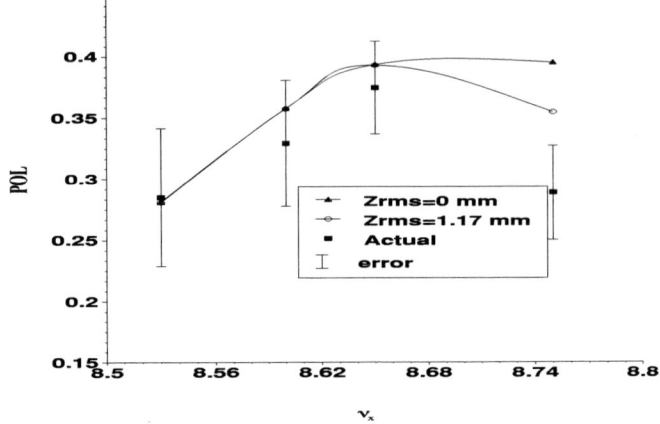

FIGURE 3. $24 + v$ resonance crossing at fixed 8.8 vertical tune, as a function of horizontal tune measured at $G\gamma = 34.5$ flattop plotted with DEPOL predictions assuming initial polarization at $G\gamma = 30.5$ to be 43%. Plots with and without added closed orbit distortions $Z_{rms} = 1.17$mm and $X_{rms} = 0.079$mm. Emittances in the vertical were 23π mm-mrad.

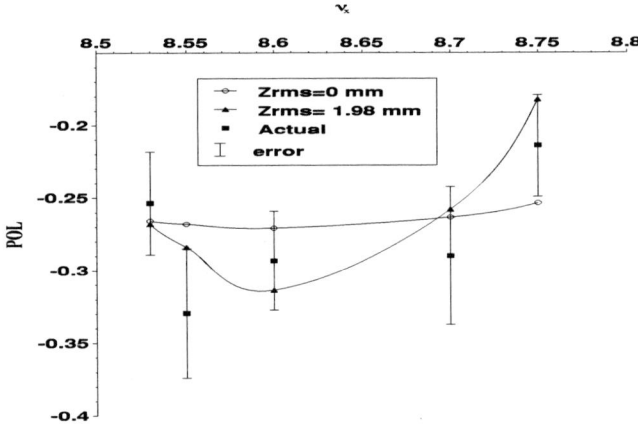

FIGURE 4. $48 - v$ resonance crossing at fixed 8.8 vertical tune, as a function of horizontal tune measured at $G\gamma = 41.5$ flattop, plotted with DEPOL predictions assuming initial polarization at $G\gamma = 34.5$ to be 32%. Curves are with and without added closed orbit distortions $Z_{rms} = 1.98$mm and $X_{rms} = 0.13$mm. Emittances in the vertical were 23π mm-mrad.

In Fig. 5 the bare AGS and MAD tune calculations are shown to be in good agreement.

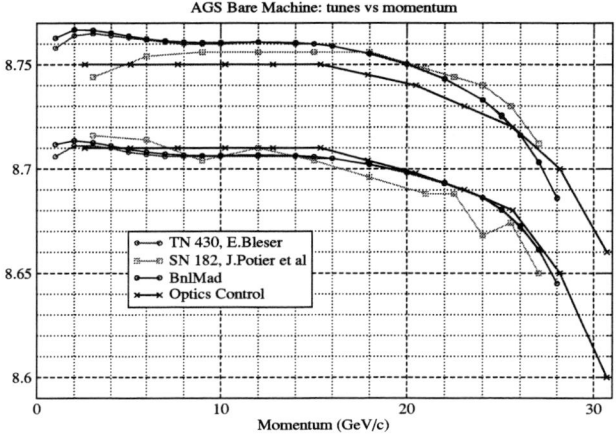

FIGURE 5. AGS bare machine tunes are plotted versus momentum. Here line labeled SN 182 represent the measured tunes performed by J. Poiter, et. al., the line labeled TN43 and BnlMad represent MAD calculations from two lattices and the line labeled optics control represents the current tune control program's predictions [9].

In Fig. 2 we can see a plot of polarization at $G\gamma = 18.5$ versus the horizontal tune at the $24 - \nu$ resonance crossing. We found that attempts to correct the 9th orbital harmonic while on the $G\gamma = 18.5$ flattop ramp produced a large polarization response. These corrections caused a variation in polarization from 55.55% to 50%. The sensitivity of polarization measurements to orbit corrections support speculation that this extra loss is due to the feed down from the sextupole fields. So we see that the addition of closed orbit errors can effect the strength and structure of intrinsic resonances. Another possible explanation is the existence of uncorrected imperfection resonances. In fact there is a precedence for this supposition, since earlier on during this experiment the existence of a corrected imperfection resonance between $G\gamma = 30.5$ and $G\gamma = 34.5$ prompted a modification of the partial snake strength and the construction of a modified snake ramping routine. In this routine the snake was kept at 3% from injection to $G\gamma = 12 + \nu$ and ramped up to 5% to $G\gamma = 46.5$. Possibly a combination of effects could explain the deviation from the expected DEPOL values, since vertical closed orbit errors could introduce both uncorrected imperfection resonances and an intrinsic resonance effect through the feed down of sextupole fields in the vertical plane. In Figs. 3 to 4 we see further evidence which the role an uncorrected closed orbit could be playing. Across all of these resonances it was found that the introduction of a vertical closed orbit error $z_{rms} > 1$ mm could improve the fit of DEPOL calculations through the sextupole feed down. The best estimate of closed orbit errors in the AGS range from $z_{rms} = 1$ to 2 mm . In future runs, much consideration should be given to correct the orbit since an uncorrected orbit can both hamper the performance of the partial snake and modify the structure of the intrinsic resonances.

PROPOSAL FOR THE ADDITION OF NEW QUADRUPOLE FAMILY

If we had the freedom to add an additional family of quadrupoles to the AGS lattice, we should be able to pick the location where the the imaginary or real part of the resonance could be controlled. So for example if we choose the 15th position which corresponds to $22.5°$ then we can control the real part for all $12n \pm \upsilon$ with n even and the imaginary part for odd n. In principle, provided that there are no other limitations, we can suppress these components of the resonance strength to an arbitrary degree. This is exactly what was published in [10] on suppressing intrinsic spin harmonics. These results are re-confirmed in Figs. 6 - 8 where the $24 \pm \upsilon$ and $48 - \upsilon$ have been successfully suppressed using a family of 12 quadrupoles of identical design as the existing tune quadrupoles.

The compensation scheme for each individual weak intrinsic resonance requires a different quadrupole setting. Since the resonances frequency of the ac dipole system must be set manually before a run, the vertical betatron tune is also fixed at strong intrinsic resonance. If we want to simultaneously suppress weak intrinsic resonances and operate the ac dipole at strong intrinsic resonances, we have to switch back and forth between different quadrupole settings. For most of the intrinsic resonances this should not be a problem [11].

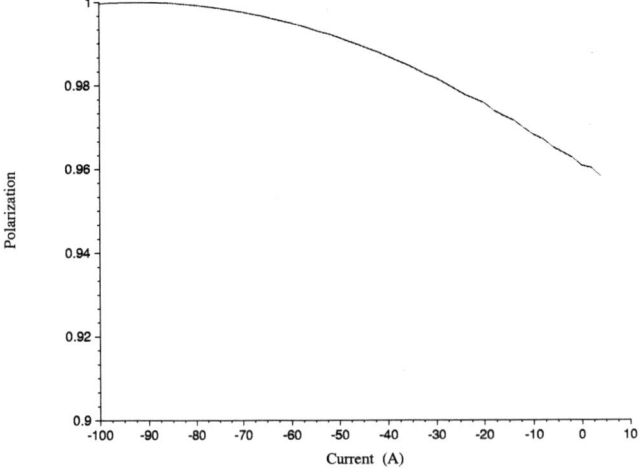

FIGURE 6. Degree of preserved polarization after crossing the $24 - \nu_z$ resonance with fixed vertical tune and horizontal tune ($\nu_z = 8.7$, $\nu_x = 8.8$). Scanning currents for a set of hypothetical quadrupoles at the 15th position in the each superperiod. The vertical rms emittance for a 10π mm-mrad beam was used. The acceleration rate was $\alpha = 2.4 \times 10^{-5}$ and snake strength set to 3%.

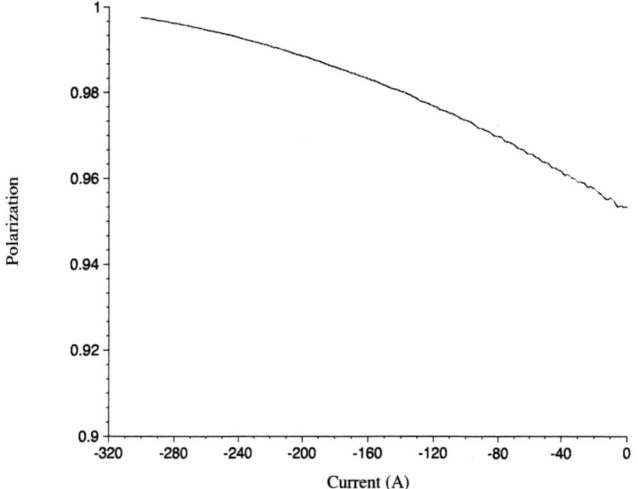

FIGURE 7. Degree of preserved polarization after crossing the $24 + v_z$ resonance with fixed vertical tune and horizontal tune ($v_z = 8.95$, $v_x = 8.6$). Scanning currents for a set of hypothetical quadrupoles at the 15th position in the each superperiod. The vertical rms emittance for a 10π mm-mrad beam was used. The acceleration rate was $\alpha = 2.4 \times 10^{-5}$ and snake strength set to 5%.

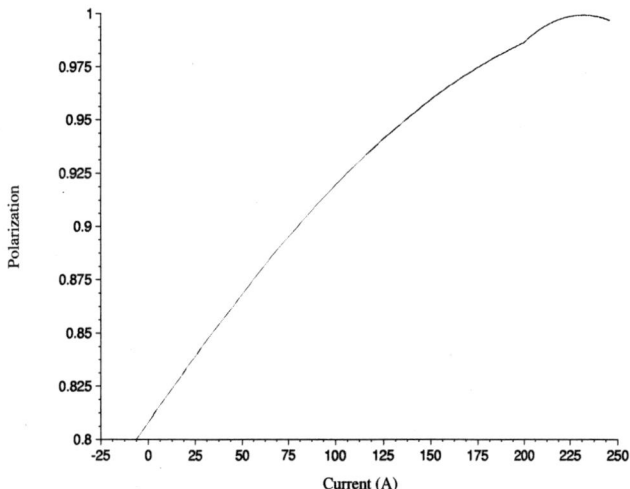

FIGURE 8. Degree of preserved polarization after crossing the $48 - v_z$ resonance with fixed vertical tune and horizontal tune ($v_z = 8.7$, $v_x = 8.6$). Scanning currents for a set of hypothetical quadrupoles at the 15th position in the each superperiod. The vertical rms emittance for a 10π mm-mrad beam was used. The acceleration rate was $\alpha = 2.4 \times 10^{-5}$ and snake strength set to 5%.

Unfortunately we are limited by the inductance of the existing tune quadrupoles. The required compensation of weak intrinsic resonances also result in betatron tunes close to betatron resonances. We will probably have to back away from these betatron resonances.

CONCLUSION

Our experimental results suggest that while it is clearly possible to suppress the remaining weak intrinsic resonances using the technique of spin matching, control of the vertical closed orbit distortions is critical. A bad orbit can distort the structure and enhance the magnitude of the depolarizing intrinsic resonance. Orbit correction is vital not only to minimize the intrinsic resonances but also to allow the operation of the partial solenoidal snake at a lower strength. This is because with a better orbit the strength of the imperfection resonances will be reduced and as such require a weaker snake to compensate. This in turn will lower the overall coupling in the machine and the strength of the coupled spin resonances. Once a good orbit is achieved then each weak intrinsic resonance can be minimized partially using the existing tune quadrupoles or totally by installing a family of low inductance quadrupoles in the 15th straight section. Using this method the final polarization can be improved by a factor of about 1.14.

REFERENCES

1. Huang H. et al., *Phys. Rev. Lett.* **73**, 2982 (1994).
2. Bai M. et al., *Phys. Rev.* **E56**, 6002 (1997).
3. Lee S.Y., Spin Dynamics and Snakes in Synchrotrons, World Scientific Singapore, ISBN 981-02-2805-8, pp. 25-42, and 93 (1997).
4. Courant E.D. and Ruth R.D., The Acceleration of Polarized Protons in Circular Accelerators, BNL 51270 (1980).
5. Lehrach A., PhD-thesis Universität Bonn (1997), Jülich Report Juel-3501, ISSN 0944-2952 (1998).
6. Steier C. and Husmann D., Correction of Depolarizing Resonances in ELSA, Proc. of the European Particle Accelerator Conference EPAC98, Barcelona, pp. 1033-1035 (1998).
7. Lehrach A., Gebel R., Maier R., Prasuhn D., Stockhorst H., *Nucl. Instrum. and Methods* **A 439**, 26 (2000).
8. Ranjbar V.H. et al., Mapping out the full spin resonance structure of RHIC, Proc. of the Particle Accelerator Conference PAC01, Chicago, pp. 3177-3179 (2001).
9. Chart obtained from Brown K., Brookhaven National Laboratory.
10. Lehrach A. et al., Suppressing intrinsic spin harmonics at the AGS, Accelerator Physics Notes C-A/AP#11, C-A Dept., Brookhaven National Laboratory (2000).
11. Ranjbar V.H. et al., Spin Matching for Weak Intrinsic Resonances in the AGS, Accelerator Physics Notes C-A/AP#41, C-A Dept., Brookhaven National Laboratory (2001).

The AGS CNI Polarimeter[1]

G. Bunce*[†], I.G. Alekseev**, A. Bravar*, S. Dhawan[‡], H. Huang*,
V. Hughes[‡], G. Igo[§], O. Jinnouchi[¶], V. Kanavets**, K. Kurita[†,#], Z. Li*,
W. Lozowski[††], W.W. MacKay*, Y. Makdisi*, S. Rescia*, T. Roser*,
D.N. Svirida**, C. Whitten[§], and J. Wood[§]

*Brookhaven National Laboratory, Upton, NY, 11973 USA
[†]RIKEN BNL Research Center, Upton, NY, 11973 USA
**ITEP, Moscow, Russia
[‡]Yale, New Haven, CT 06520 USA
[§]UCLA, Los Angeles, CA 90095 USA
[¶]RIKEN, Wako, Japan
[#]Rikkyo University, Tokyo, Japan
[††]Indiana University, Bloomington, IN 47405 USA

Abstract. A new polarimeter is being installed in the Brookhaven AGS, based on the successful proton-carbon polarimeters in RHIC. The polarimeter will measure the left-right asymmetry for proton—carbon elastic scattering in the Coulomb-nuclear interference (CNI) region, for vertically polarized protons in the AGS. The polarimeter offers a much higher figure of merit than the existing AGS polarimeter which is based on larger angle proton—proton elastic scattering. We expect to measure the polarization in the AGS with a single or a few acceleration cycles. We also plan to measure the polarization in 2 ms bins during the AGS acceleration ramp. Multiple ramps will be necessary, probably over 30 minutes to an hour.

INTRODUCTION

The RHIC polarimeters scatter the polarized proton beam from an ultra-thin carbon target that is inserted into the beam for the measurement. Elastic scattering is selected in the range of $-t=0.003$ to 0.02 $(GeV/c)^2$, where we observe only the recoil carbon ions scattered to near $90°$ from the beam. Time of flight, compared to the time that the rf-bunched beam in RHIC passed through the target, and energy are measured with silicon strip detectors. The relationship between energy and velocity of the recoil particles is used to cleanly identify carbon scatters. The scattering is sensitive to the beam polarization through a left right asymmetry for vertically polarized protons. This asymmetry is due to the interference of an electromagnetic spin-flip scattering amplitude, proportional to the proton's anomalous magnetic moment, with a hadronic non-flip amplitude, proportional to the square root of the hadronic cross section. The maximum analyzing power is expected for $-t=0.003$ (GeV/c) and is $A_N=0.04$. We observe a lower analyzing power, interpreted to be due to a negative hadronic spin-flip contribution to A_N. The average analyzing power at RHIC is $A_N=0.013$, for the range

---

[1] Supported by the U.S. Department of Energy and RIKEN, Japan.

−t=0.007 to 0.03 (GeV/c)^2. The RHIC polarimeters measure the beam polarization in less than a minute, as shown in Fig. 1. The RHIC CNI polarimeters are described in [1].

The present AGS polarimeter is based on proton—proton elastic scattering at an observed peak of analyzing power, for −t=0.15 (GeV/c)^2 ([2], Appendix). This peak value is observed (using polarized targets with known polarization) to fall with incident beam momentum as $1/p_{incident}$. The analyzing power at the RHIC injection energy of 24 GeV is A_N=0.01. A measurement at this energy with the present AGS polarimeter takes about 30 minutes. The energy where the CNI polarimeter figure of merit, $\sigma \times A_N^2$, becomes larger than that for the p—p polarimeter is around 5 GeV.

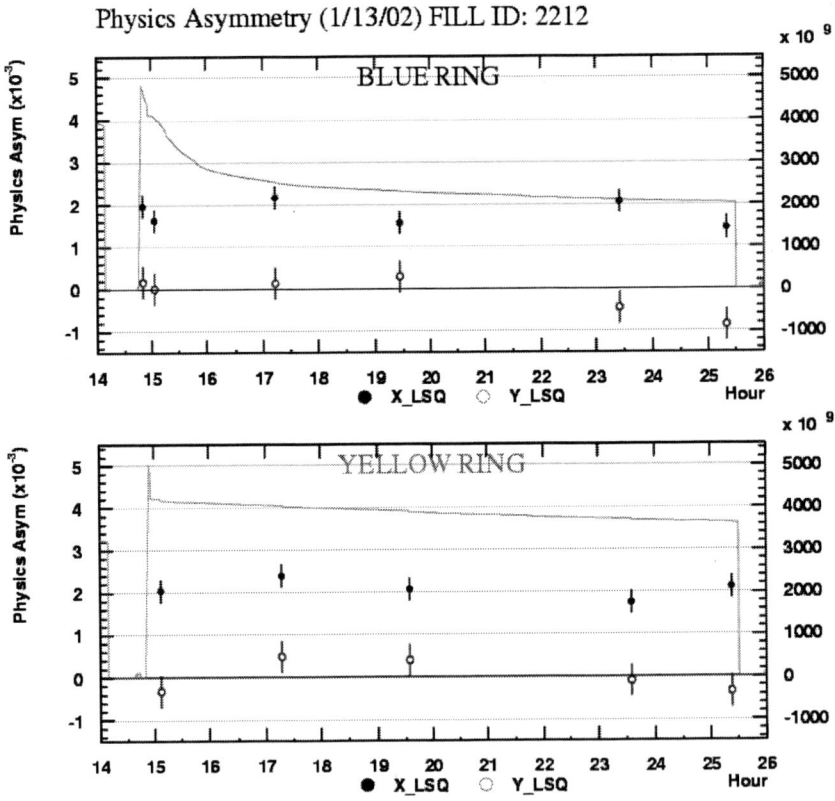

Figure 1. Measured physics asymmetries along with the polarized proton intensity as a function of time, for a typical fill. The figures show the blue ring (top), and the yellow ring. The closed points show the horizontal asymmetry, and the open points the vertical asymmetry, with the scale on the left axis. For a vertical polarization, only a horizontal asymmetry is expected. The first measurement in the blue ring was at 24 GeV, and all other measurements are at 100 GeV. The thin lines show the beam intensity, referring to the right axes. See [1] for details.

The new AGS CNI polarimeter is expected to provide much shorter measurement time, compared to the p—p polarimeter, and measurements on the acceleration ramp.

PLAN AND STATUS OF THE AGS CNI POLARIMETER

The AGS application required several changes to the RHIC polarimeter design. We want to measure the polarization from AGS injection, momentum 2.3 GeV/c, up to the extraction momentum for RHIC, 24 GeV/c, and measure on the acceleration ramp. The low energy measurements, where the transverse beam size is large, require longer carbon targets, 5 cm in place of the 3 cm targets in RHIC. We also require wider targets so that we can accumulate sufficient counts in a reasonable time for the measurements on the ramp. Bill Lozowski [3] did additional development on the ultra-thin carbon targets, and we now have 5 cm long targets, 150 angstroms thick, with various widths up to 600 μm. (The RHIC targets are as narrow as 3.5 μm.) We also increased the distance to the silicon detectors from the target to 25 cm, compared to 15 cm at RHIC (refer to [1], Fig. 1). This was necessary to improve the time of flight discrimination of the recoil carbon ions, because of the longer rf bunches in the AGS. We have also mounted only two silicon detectors, left and right. Detectors at 45° were not installed because we will not measure radial polarization at this time. We have decided to use 6 bunches in the AGS with alternating polarization sign, both to increase the data collection rate (the AGS normally uses one bunch when polarized protons are injected into RHIC), and to include both polarization signs for each AGS cycle, as we do in RHIC. Finally, we needed to modify the read-out to measure in 2 ms bins on the acceleration ramp. This read-out approach was developed for the RHIC CNI polarimeters, and is described next.

Fig. 2 shows the time of flight vs. energy for a single silicon strip at RHIC. This plot is obtained directly from the wave-form digitizer (WFD) used to read out the silicon [4]. Flash ADC energy measurements are made by the WFD each 2.5 ns, and an on-board FPGA chip finds the maximum, determining the energy of the maximum and the time of flight. For the 2001/2 RHIC run, a look up table in the WFD selected events which fell in the carbon window, as shown in the figure. Scalers in the WFD then kept track of the number of events collected for each bunch, for each strip. The bunches alternate in polarization, so that left-right asymmetries are collected in the same fill for up and down polarization.

For the AGS CNI polarimeter, we have added binning vs. time in the AGS acceleration ramp, and we will also keep information for each event, taking advantage of increased memory in the WFDs. The spacing of the spin resonances on the acceleration ramp is 10 ms and less, corresponding to 500 MeV steps in energy for the imperfection spin resonances, with additional intrinsic resonances in between. At many of the resonances the spin flips, so that it is important that the binning avoid overlapping two resonances. We have chosen 2 ms bins. Work on this change is in progress.

We will use the p—p polarimeter to cross-calibrate the CNI polarimeter at several energies. This polarimeter requires debunched beam on flattop, vs. the 6 bunches with alternating polarization sign that we will normally use for the CNI polarimeter. The p—p polarimeter also uses p—p elastic scattering, and normally requires a hydrocarbon target (nylon is used). However, a nylon target does not survive the present high intensity polarized beam. We can use p—p quasielastic scattering from a carbon

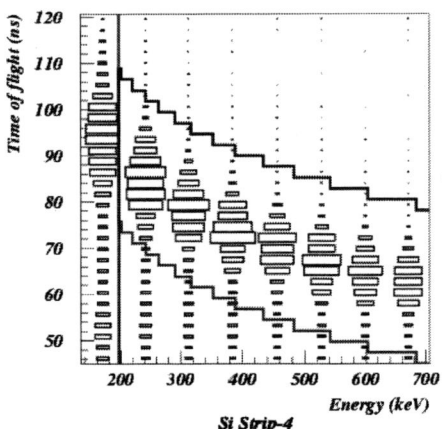

Figure 2. Time of flight vs. energy for one silicon strip of the RHIC polarimeter. The intense band corresponds to the recoil carbon ions. This scatterplot is stored in the wave form digitizer readout.

target (this is the setup used last year for the p—p polarimeter), which has roughly half the analyzing power of p—p elastics. We can also greatly reduce the beam intensity and install a nylon target, if necessary.

We will also investigate whether we can use the new AGS CNI polarimeter and the p—p polarimeter to improve the calibration of the RHIC CNI polarimeters at injection. To do this, we would need to install a nylon target for the p—p polarimeter. Here we would expect improved systematic errors over the initial experiment where we developed the CNI polarimetry method [5], which also provides the present calibration of the RHIC polarimeters.

The CNI polarimeter is installed in the AGS, and work is proceeding on handling noise, and the new WFD coding is being installed. We expect to be ready to study the AGS polarization by late January 2003.

REFERENCES

1. O. Jinnouchi *et al.*, "RHIC pC CNI Polarimeter: Status and Performance from the First Collider Run", *SPIN 2002 Proceedings*, BNL Sept. 2002, to be published by AIP.
2. C.E. Allgower *et al.*, *Phys. Rev.* **D65**, 092008 (2002).
3. W.R. Lozowski, private communication; W.R. Lozowski and J.D. Hudson, *NIM* **A334**, 173 (1993).
4. S. Dhawan, private communication; I.G. Alekseev *et al.*, "Wave-form Digitizer for RHIC Polarimetry", *SPIN 2002 Proceedings*, BNL Sept. 2002, to be published by AIP.
5. J. Tojo *et al.*, *Phys. Rev. Lett.* **89**, 052302 (2002).

A No-Depolarization Theorem for Rotator-Aided Resonance Crossing

Dennis W. Sivers

Portland Physics Institute, Portland, Oregon 97201
and
Spin Physics Center, University of Michigan, Ann Arbor, MI 48109-1120

Abstract. An rf frequency rotator magnet provides a useful tool for manipulating particle spins in any accelerator or storage ring with polarized beams. This note briefly demonstrates the general idea of the rotator-aided crossing of spin resonance and sketches the proof of a general theorem about the process. The important question of whether this technique can be useful for a specific spin resonance at a particular accelerator involves a detailed analysis of the impact of the rotator on a range of machine parameters.

The use of coherent rotator magnets for the manipulation of particle spins in accelerators and storage rings has a venerable history. For example, in a series of experiments at IUCF [1], a group led by A. Krisch has conclusively demonstrated the value of both rf-dipole and rf-solenoid rotators. It is particularly important to note the results of the Brookhaven accelerator group reported at this meeting [2], involving the use of an rf-dipole magnet to aid in the crossing of intermediate-strength intrinsic resonances in the AGS.

Instead of dealing with a specific situation, it is constructive to first consider a general theorem involving the use of rotators. For this purpose, a rotator will be defined as a device which produces a localized magnetic field intersecting the beam as given by,

$$\vec{B}_i = \hat{e}_i \left(C_m \cos v_m t \right) \qquad (1)$$

In this context it will be assumed that the magnet parameters can be adjusted independently and are controllable on a time scale that is long compared to v_m^{-1}. That is, we can write

$$C_m = C_m(t) \qquad (2)$$

$$v_m = v_m(t) \qquad (3)$$

and vary these parameters to aid in resonance crossing. It is important for this discussion that it is possible to vary C_m over a sufficient range to produce an induced spin resonance capable of generating a spin flip.

We can think of a spin resonance as an operator that impacts the polarization of a beam when the spin tune, $v_s(t)$, passes the resonance frequency. The strength of a resonance is important. If a spin resonance is sufficiently weak, it is possible to cross the resonance rapidly with little or no depolarization. Conversely, if it is sufficiently strong, it is possible to use the Froissart-Stora [3] technique of adiabatic crossing to induce a spin flip. Measuring the beam polarization before and after crossing, we label the two limits

$$P_f / P_i \cong 1 \tag{4}$$

$$P_f / P_i \cong -1 \tag{5}$$

as "controlled" crossings since they do not destroy beam polarization. Those resonances which are dangerous to polarization are of intermediate strength and do not allow "controlled" crossings.

A rotator magnet operating as described above creates an induced spin resonance at $v_m(t)$ with a strength that can be varied from weak to strong depending on the value of $C_m(t)$. With these preliminaries, we can formulate a no-depolarization condition with the following theorem.

Theorem. Assume that a spin resonance, v_{res}, divides the plane $\{v \in [v_{min}, v_{max}]$ and $t \in [t_{min}, t_{max}]\}$ into two disjoint regions so that it is not possible to form a path, $v_s(t)$, that passes from $v_s(t_{min}) \leq v_{res}$ to $v_s(t_{max}) \geq v_{res}$ without crossing the resonance. By adding an additional, induced, resonance at $v_m(t)$ to the system, the $\{v, t\}$ plane is separated by the two resonances into three disjoint regions. Varying the magnet parameters $C_m(t)$ and $v_m(t)$ for the induced resonance allows a path $v_s(t)$ that navigates from $v_s(t_{min}) \leq v_{res}$ to $v_s(t_{max}) \geq v_{res}$ with either no resonance crossings or two "controlled" resonance crossings.

The proof of this simple theorem utilizes quantum-mechanical operator mixing and the fact that the interference of the operators prevents an overlap of two resonances so that the $\{v, t\}$ plane is cut by the two resonance lines as indicated in Fig. 1. In addition, in regions where operator mixing is important, the interference effects that mix the operators assure that both resonances will have similar strength. Varying the magnet parameters therefore allows control location of the induced resonance and of the strength of both resonances over a finite region of the $\{v, t\}$ plane. Coordination of the acceleration path $v_s(t)$ with the variation of the magnet parameters therefore allows a path which either avoids the "mixed" resonances or executes two "controlled" crossings.

To convert this from a simple mathematical plane-cutting theorem to a physics "no depolarization" theorem requires a considerable understanding of the impact of the operation of the rotator magnet on other accelerator parameters. The theorem cannot guarantee that the technique will be practical. Unlike the operation of a Siberian Snake [4], the use of a coherent rotator magnet merely aids in the crossing of a resonance; it does not eliminate a class of resonances. Application of the technique requires testing and polarization measurements to achieve optimal results. Because of the time required to perform a polarization measurement, this can be a serious obstacle. However, once a successful resonance-crossing strategy has been determined, it can be successfully repeated without further input from polarization measurements. The experiments at IUCF have shown that adiabatic spin flips through controlled resonance crossings require tuning efforts. Avoiding a resonance crossing does not guarantee that there will be no depolarization. Because of spin-tune "spread" and other effects, operating a beam near a resonance for a finite amount of time can lead to depolarization just like a resonance crossing. The results of the Brookhaven group justify the use of resonance crossings.[2] The techniques of varying the magnet parameters outlined here merit further study but do not guarantee results.

ACKNOWLEDGEMENTS

The co-chairs of this conference, A. Krisch and T. Roser, created a collaborative atmosphere that was very encouraging to new ideas, and the hospitality of the Spin Physics Center at Michigan was very nourishing. As a result, the author received considerable aid from suggestions by Ya. Derbenev, L. Teng and M. Bai.

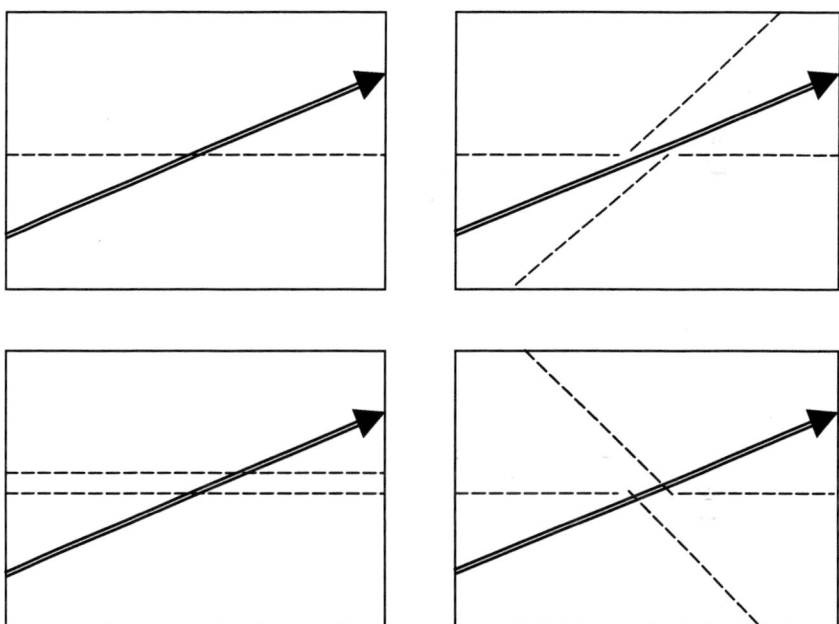

FIGURE 1. Four sketches of the tune vs. time plane with different resonance configurations. Top left, one resonance. Top right, induced resonance with increasing frequency. Bottom left, induced resonance with fixed frequency close to intrinsic resonance. Bottom right, induced resonance with decreasing frequency. Arrows indicate possible path for spin tune of ring.

REFERENCES

[1.] See, for example, B. Blinov et al., Phys. Rev. Lett. **81**, 2906 (1998); V. Anferov et al., Phys Rev ST-AB **3**, 041001 (2000); B. Blinov et al., Phys Rev ST-AB **3**, 104001 (2000).
[2.] M. Bai, these proceedings. See, also, M. Bai and T. Roser, "Crossing a coupling spin resonance with an RF dipole" in *Proceedings of the 14th Int'l Spin Physics Symposium*, Osaka, Japan, October 2000, edited by K. Hatanaka et al., AIP Conference Proceedings **570**, New York (2001), p. 741.
[3.] M. Froissart and R. Stora, *Nucl. Instrum. Methods* **7**, 297 (1960).
[4.] Ya. Derbenev and A. Kondratenko, Sov. Phys. JETP **37**, 968 (1973).

Spin Matching from AGS to RHIC[1]

W. W. MacKay, and N. Tsoupas

Brookhaven National Laboratory, Upton, NY 11973, USA.

Abstract. With a partial Siberian snake in the AGS and transport lines with interspersed horizontal and vertical bends, the incoming spin direction at the injection points of both the collider rings is not likely to match the ideal vertical stable spin direction of RHIC which has two full helical Siberian snakes per ring. In this paper we examine the matching of a polarized beam transferred from the AGS into RHIC. The present 5% partial solenoidal snake as well as a proposed 20% superconducting helical are considered for the AGS. Solutions with retuned snakes in RHIC to better match the incoming beam have been found.

Ideally in a flat ring without snakes, we should expect the stable spin direction $\vec{n}_0$ of the closed orbit to be vertical (except at spin resonances). The collider accelerator complex[1][2] at BNL consists of a polarized proton source followed by a linac, the Booster ring, the AGS ring and the two collider rings, not to mention the connecting transport lines. At injection ($G\gamma = 2.18$) and extraction ($G\gamma = 4.7$) in the Booster ring, the stable spin direction is vertical. At present the AGS has a warm solenoidal partial Siberian snake which rotates the stable direction away from the vertical direction. When the partial snake in the AGS is operated with a $\sim 5\%$ rotation (9°), the stable spin direction is tilted 4.5° away from the vertical. Each of the collider rings has a pair of full (180°-rotation) superconducting helical Siberian snakes to fix the spin tune exactly at 0.5, independent of beam energy. With the snakes the injection points of the collider rings have stable spin directions which are vertical. Injection and extraction in the Booster and AGS rings happen in horizontal plane, but for RHIC we inject vertically.

The AGS is about 1.7 m higher than the RHIC rings, so in the AGS-to-RHIC transfer line (ATR) there are vertical bends interspersed with the normal horizontal bends. Due to the partial snake in the AGS and the vertical bends in the ATR, spin matching from the AGS into RHIC is not perfect. By retuning the RHIC snakes for injection, we can improve the spin transfer from the AGS. In this paper we examine the spin matching from the AGS into RHIC with the present 5% snake and with a proposed stronger superconducting helical 20% snake.

PRELIMINARIES

First we define our coordinates with positive angles for clockwise bends as in the Blue ring of RHIC as shown in Figure 1. The Pauli matrices for the three directions are defined

---

[1] Work performed under the auspices of the U. S. Department of Energy and RIKEN of Japan.

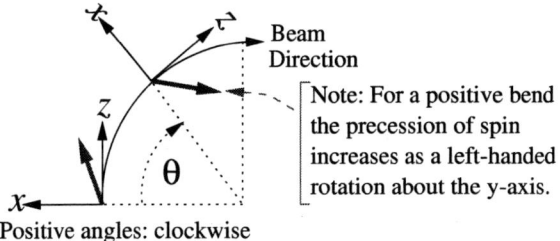

FIGURE 1. For the local coordinate system traveling with the beam, the z-axis points along the direction of the beam, y-axis points vertically (out of the page), and the x-axis points to the left thus forming a right-handed system. Clockwise bends (to the right) have a positive bend angle. The projection of the polarization in the horizontal plane is measured relative to the local z-axis.

as

$$\sigma_x = \begin{pmatrix} 0 & 1 \\ 1 & 0 \end{pmatrix}, \quad \sigma_y = \begin{pmatrix} 0 & -i \\ i & 0 \end{pmatrix}, \quad \text{and} \quad \sigma_z = \begin{pmatrix} 1 & 0 \\ 0 & -1 \end{pmatrix}, \qquad (1)$$

so that a left-handed spin rotation about an axis $\hat{n}$ by an angle θ is then given by the 2×2 spinor rotation matrix

$$\begin{aligned} \mathbf{R}_{\hat{n}}(\theta) &= e^{i\hat{n}\cdot\vec{\sigma}\theta/2} = \mathbf{I}\cos\tfrac{\theta}{2} + i\hat{n}\cdot\vec{\sigma}\sin\tfrac{\theta}{2} \\ &= \begin{pmatrix} \cos\tfrac{\theta}{2} + in_z\sin\tfrac{\theta}{2} & (n_y + in_x)\sin\tfrac{\theta}{2} \\ (-n_y + in_x)\sin\tfrac{\theta}{2} & \cos\tfrac{\theta}{2} - in_z\sin\tfrac{\theta}{2} \end{pmatrix}. \end{aligned} \qquad (2)$$

Note these conventions are different from S. Y. Lee's book[1] and coincide more closely to the usual conventions of quantum mechanics and high energy physics [2]. The defined coordinate systems vary from the Booster (clockwise) to the AGS (counterclockwise) through the ATR (AGS to RHIC transfer line) with a clockwise system to the clockwise Blue ring, and the counterclockwise Yellow ring. (Note that in the Yellow ring the s-coordinate is defined parallel the Blue ring [3], i.e. opposite to the direction of the beam.) Needless to say things can get very confusing in switching coordinates systems and rotation directions throughout the transport of the beam.

With all this in mind for this paper, we should note that in the AGS and Yellow ring our convention has $+x$ pointing toward the center of the rings. In the Booster and the Blue ring $+x$ points away from the center of the rings. With respect to spin rotation this gives:

$$\mathbf{R}_{\hat{x}}(\pi/2) \quad \text{rotating } \hat{z} \text{ into } \hat{y},$$
$$\mathbf{R}_{\hat{y}}(\pi/2) \quad \text{rotating } \hat{z} \text{ into } -\hat{x}, \text{ and}$$
$$\mathbf{R}_{\hat{z}}(\pi/2) \quad \text{rotating } \hat{y} \text{ into } \hat{x}.$$

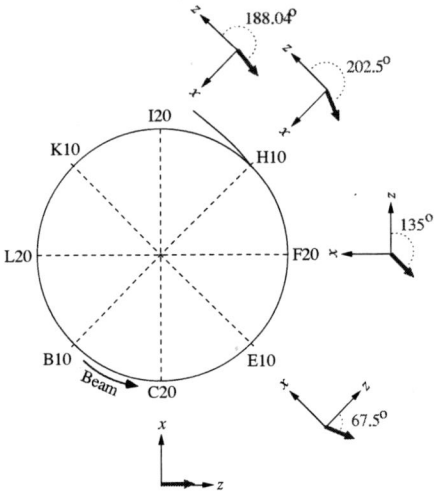

FIGURE 2. Basic geometry of the AGS spin precession with a 20% partial snake in the I20 straight section at extraction energy ($G\gamma = 46.5$). For the superconducting helical snake being considered, the snake's rotation axis is longitudinal as in a solenoid.

AGS EXTRACTION

In the AGS, the beam goes around in the counterclockwise direction as shown in Figure 2. The extraction location is in the H10 straight section, and the new snake will most likely be placed in the I20 straight section. The azimuthal angle from H10 to I20 is

$$-\tfrac{3}{24} \times 2\pi = -\tfrac{\pi}{4}, \tag{3}$$

and the angle from the snake back around to H10 is

$$-\left[2\pi - \tfrac{\pi}{4}\right] = -\tfrac{7\pi}{4}. \tag{4}$$

The corresponding amount of spin precession in these arcs is then

$$\eta_1 = -\tfrac{\pi G\gamma}{4}, \quad \text{and} \quad \eta_2 = -\tfrac{7\pi G\gamma}{4}. \tag{5}$$

The beam in the AGS goes around in a counterclockwise direction, so we have the 1-turn spin rotation map

$$\begin{aligned}
\mathbf{M} &= \begin{pmatrix} \cos\tfrac{\eta_2}{2} & \sin\tfrac{\eta_2}{2} \\ -\sin\tfrac{\eta_2}{2} & \cos\tfrac{\eta_2}{2} \end{pmatrix} \begin{pmatrix} \exp\left(i\tfrac{\mu}{2}\right) & 0 \\ 0 & \exp\left(-i\tfrac{\mu}{2}\right) \end{pmatrix} \begin{pmatrix} \cos\tfrac{\eta_1}{2} & \sin\tfrac{\eta_1}{2} \\ -\sin\tfrac{\eta_1}{2} & \cos\tfrac{\eta_1}{2} \end{pmatrix} \\
&= \mathbf{I}\cos\tfrac{\mu}{2}\cos\tfrac{\eta_1+\eta_2}{2} + i\sin\tfrac{\mu}{2}\sin\tfrac{\eta_1-\eta_2}{2}\sigma_x \\
&\quad + i\cos\tfrac{\mu}{2}\sin\tfrac{\eta_1+\eta_2}{2}\sigma_y + i\sin\tfrac{\mu}{2}\cos\tfrac{\eta_1-\eta_2}{2}\sigma_z.
\end{aligned} \tag{6}$$

The fractional tune may be obtained from the trace:

$$2\cos(\pi\nu) = 2\cos\tfrac{\mu}{2}\cos\tfrac{\eta_1+\eta_2}{2}. \tag{7}$$

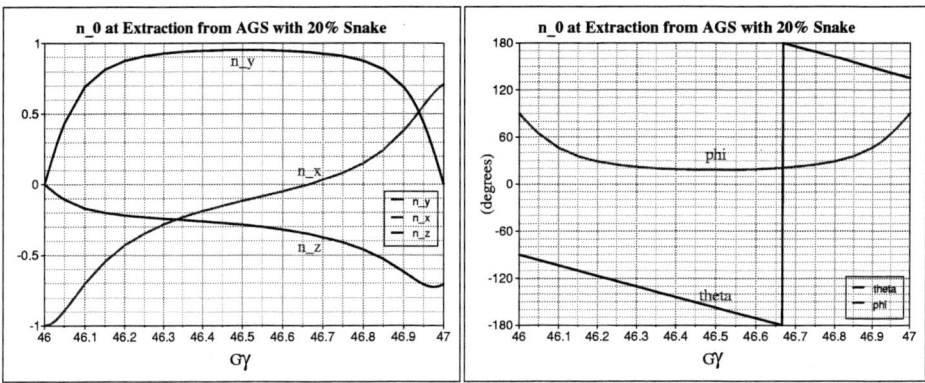

FIGURE 3. Plot of $\vec{n}_0$ at extraction energy at the H10 straight section. θ is the angle of the $\hat{n}_0$ vector away from the vertical, and ϕ is the angle between the z-axis and the projection of the $\hat{n}_0$ vector on the horizontal plane.

The stable spin direction at the H10 straight section may be easily obtained from Eq. (6):

$$\hat{n}_0 = \frac{1}{\sin(\pi\nu)} \begin{pmatrix} \sin\frac{\mu}{2}\sin\frac{\eta_1-\eta_2}{2} \\ \cos\frac{\mu}{2}\sin\frac{\eta_1+\eta_2}{2} \\ \sin\frac{\mu}{2}\cos\frac{\eta_1-\eta_2}{2} \end{pmatrix}. \qquad (8)$$

A plot of $\hat{n}_0$ versus $G\gamma$ at H10 shows the dependence on energy of the extraction from the AGS. The vertical spin component is quite stable near the half-integer values of $G\gamma$.

TRANSFER LINE FROM AGS TO RHIC

Due to interleaved horizontal and vertical bends from the AGS extraction to the RHIC injection points, the value of the $\vec{n}_0$ of the injected beam at the RHIC injection point will vary both with energy and AGS snake setting, and may differ for each of the Blue and Yellow rings. The transfer lines are divided into four sections (See Figure 4a.): first, the U-line with two horizontal bends of 4.31° and 8°; next, the W-line which provides a horizontal bend of 20° to orient the beam to the switching magnet along a mirror symmetry axis for the RHIC rings; finally the two large arcs the X-line for injection into the clockwise Blue ring and the Y-line for injection into the counterclockwise Yellow ring. Since the planes of the AGS and RHIC rings have a 1.73 m difference in height, there are a pair of vertical bends in the W-line as shown in Figure 4b. The lattice was designed to have a full period of betatron phase advance between the two bends in order to minimize vertical dispersion for injection. Beams are injected into RHIC through vertical septum (Lambertson) magnets. The nominal vertical bends from steering magnets upstream of the septum magnet, two ring quadrupoles downstream of the septum, and the injection kicker magnets (4 modules) are interleaved with horizontal bends as shown in Figure 4c.

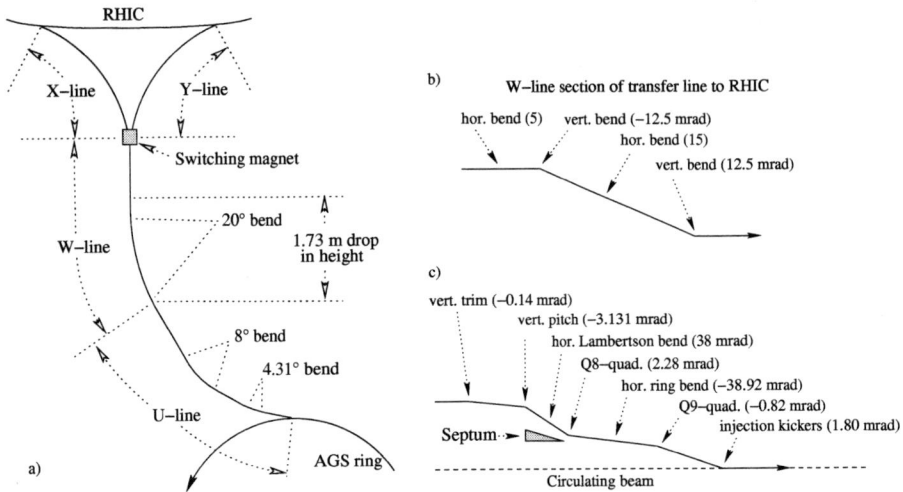

FIGURE 4. Layout of the AGS to RHIC transfer lines (ATR). a) Horizontal layout of the ATR. b) Layout of vertical bends for 1.73 m drop to the AGS. c) Vertical layout of injection to RHIC. This figure was drawn for the Yellow ring injection; the horizontal bends of 38 mrad and -38.92 mrad should reverse sign for injection into the Blue ring.

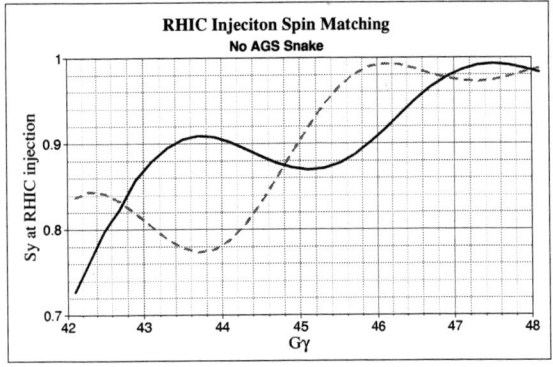

FIGURE 5. Spin matching into RHIC with 0% partial snake in the AGS. The solid (dashed) curve shows the matching into the Blue (Yellow) ring for different values of $G\gamma$.

Figure 5 shows the vertical projection of the n_0 vector at injection into RHIC rings. In the first vertical bend of the W-line, $\vec{n}_0$ tilts $\sim 30°$ away from the vertical about the radial axis, then rotates about the vertical by $G\gamma \times 15°$ in the horizontal bends, and then back by $\sim -30°$ about the radial axis in the second vertical bend.

When $\vec{n}_0$ is tilted either forward or backward at the end of the W-line, then we see a mirror symmetric precession through the X and Y-lines. This is shown in Figure 5 for $G\gamma = 48$ where there is about a net 10° tilt away from vertical in the injection region of Figure 4c.

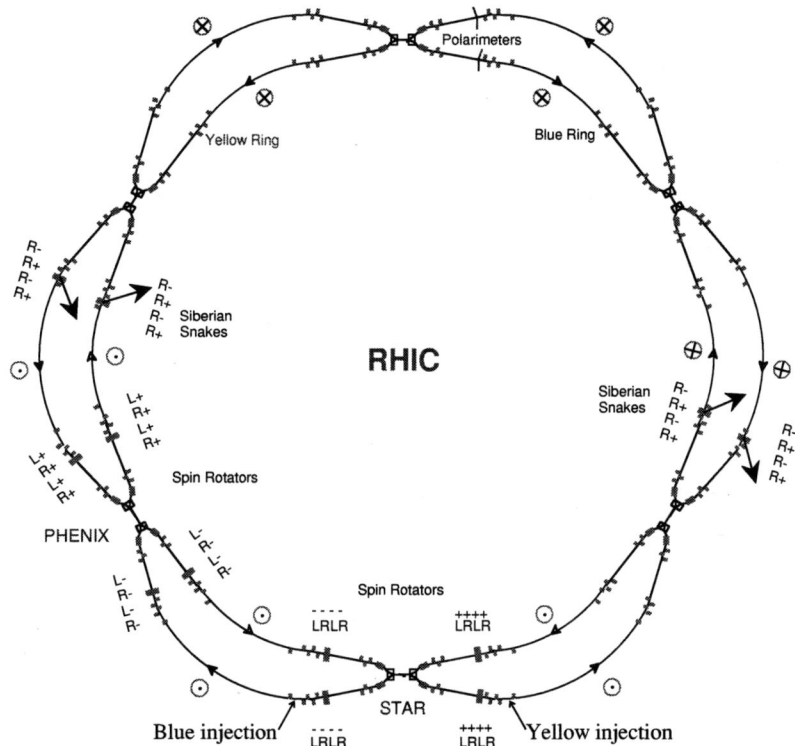

FIGURE 6. Layout of RHIC rings. In the Blue ring the beam moves in a clockwise direction, and in the Yellow ring the beam goes counterclockwise. The snakes consist of four right-handed helical dipoles with vertical fields at the ends as indicated (+ for $\vec{B}_{end}$ up, − for $\vec{B}_{end}$ down). The spin rotators consist of a combination of four right-handed (R) and left-handed (L) helical dipoles with horizontal fields as indicated (− for $\vec{B}_{end}$ pointing radially in towards the center of the rings, and + for $\vec{B}_{end}$ pointing radially outward from the center of the rings).

RHIC RINGS WITH TWO SNAKES

Figure 6 shows the placement of polarization equipment in the collider rings. The injection Lambertson magnets are located in the straight sections of dispersion suppressors at the beginning of the lower main arcs on either side of the STAR detector. The two rings are side by side horizontally and cross over at each of the six interaction regions. Two superconducting Siberian snakes (each with four helical dipoles) are located on opposite sides of each ring. There are also four superconducting rotators placed around each of the STAR and PHENIX detectors to rotate the spin locally into the longitudinal direction. A polarimeter is located in each ring just to the right of the north interaction region.

Each RHIC snake consists of four right-handed helical dipoles with a 9.5 cm diameter beam pipe. The snake is powered by two power supplies: one for the end helices in series

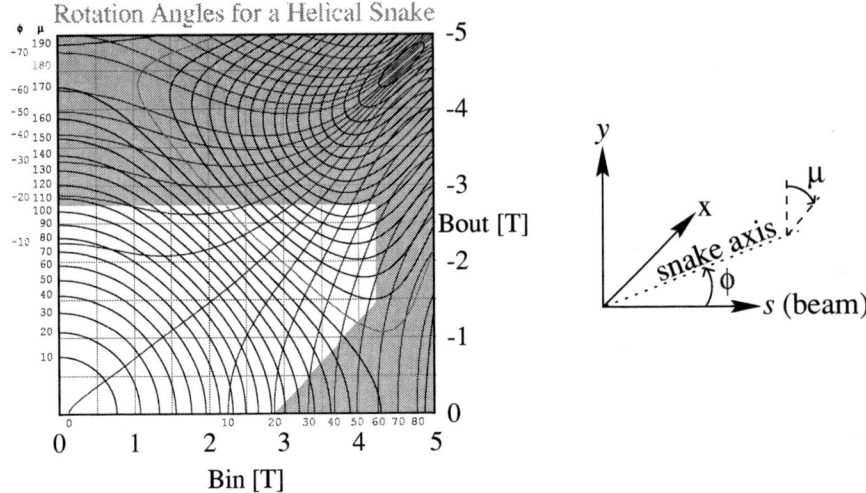

FIGURE 7. Field tuning plane for RHIC snakes. The inner pair of helices are powered to the fields B_{in} and the outer pair to B_{out}. The darker contours correspond to the rotation angle μ and the lighter contours correspond to the angle between the rotation axis and beam's direction. The shaded region corresponds to forbidden regions where the orbit excursion at injection energy reaches the aperture limit, or where the central helices would reach too high a field.

and another for the inner pair of helices also in series. The orbit excursion from one helix is canceled by having the paired helix powered with opposite polarity.

Figure 7 shows the field settings for the two pairs of helices in a given snake to obtain different amounts of rotation of the polarization. With two knobs, there is control of both the amount of rotation μ and the direction of the rotation axis ϕ. Although the superconducting helices in the snakes can operate up to 4.2 T, at injection the orbit excursions limit the operating range to the unshaded region of Figure 7. For normal operation the largest orbit excursion is about 3.1 cm vertically in the middle of the snake; this is right at the edge of the shaded contour, $(B_{in}, B_{out}) = (4\,T, 1.2\,T)$. It would be possible to reposition a snake by as much as 1 cm to allow for some modest amount of tuning at injection.

For this study we have kept the snake rotation axes at $\pm 45°$ in each ring to maintain the spin tune at 0.5. Table 1 shows the injection transfer efficiency to both collider rings with a 20% partial snake located in the I20 straight section of the AGS. Clearly the the efficiencies are better if the partial snake field is antiparallel to the beam (-20%). The first two rows show efficiencies with nominal settings for the RHIC snakes, and the third row shows an improvement if the 1st snake after the injection point has been retuned by mainly reducing the B_{out} setting to better match the incoming beam. The last two rows show that with the retuned RHIC snake, the Yellow beam would be rather insensitive to slight changes in the injection energy, whereas the Blue injection could be improved by a percent if the injection energy was lowered by $\Delta G\gamma = 0.1$.

Table 2 shows the effect of locating the partial snake at several other long straight sections in the AGS with the nominal settings for the RHIC snakes (i.e. a vertical stable

TABLE 1. Matching of $\hat{n}_0$ into RHIC for various snake settings. The calculations are for a new snake in the AGS in the I20 straight section.

		Blue			Yellow		
$G\gamma$	AGS Snake	μ_1	μ_2	Match	μ_1	μ_2	Match
46.5	-20%	180°	180°	0.982	180°	180°	0.966
46.5	+20%	180°	180°	0.824	180°	180°	0.912
46.5	-20%	172°	180°	0.985	164°	180°	0.975
46.4	-20%	172°	180°	0.996	164°	180°	0.975
46.6	-20%	172°	180°	0.965	164°	180°	0.973

TABLE 2. Matching of $\hat{n}_0$ into RHIC for different snake locations in the AGS. The last two rows show the matching with the present solenoidal snake located in the I10 straight section. Note that the injection septum and kicker are located in the L20 and A10 straight sections, respectively.

			Blue			Yellow		
AGS Location	$G\gamma$	AGS Snake	μ_1	μ_2	Match	μ_1	μ_2	Match
L10	46.5	+20%	180°	180°	0.985	180°	180°	0.987
L20	46.5	+20%	180°	180°	0.999	180°	180°	0.987
A10	46.5	+20%	180°	180°	0.998	180°	180°	0.980
A20	46.5	+20%	180°	180°	0.982	180°	180°	0.966
B10	46.5	+20%	180°	180°	0.955	180°	180°	0.949
I10	46.5	-5%	180°	180°	0.963	180°	180°	0.993
I10	46.5	+5%	180°	180°	0.923	180°	180°	0.974

spin direction). However for other partial snake strengths the matching might not be as good.

It is interesting to note that with the present solenoid snake in the AGS set to +5% at extraction, we would expect almost an 8% loss of polarization to the Blue with the partial snake field parallel to the beam. The last two rows of Table 2 show the spin injection efficiency for the present solenoidal partial snake. In fact in the 2002 polarized proton run the partial snake was parallel to the beam's direction, and we did observe a somewhat smaller polarization in the Blue ring than the Yellow ring. The efficiency for the Blue ring should be improved if we reverse the polarity of the partial snake in the next run.

CONCLUSIONS

Even though the polarity of the partial snake in the AGS should not matter for acceleration, it becomes important when we consider the efficiency for matching the spin into the collider rings. The best solution for a 20% partial snake at I10 in the AGS would have the partial snake field antiparallel to the beam and one snake retuned for smaller spin rotation in each of the RHIC rings. In the Blue ring, the snake would only need to be retuned from 180° to 172° rotation to have a 99.6% transmission of polarization from the AGS into the Blue ring. The Yellow ring would require a larger retuning of one snake to 164° of rotation; in this case the transmission would only be 97.5%. However tuning the snake to 164° would require a vertical realignment of the snake by about 0.5 cm to allow enough aperture for the orbit distortion in the snake at injection energy. Other solutions using one or two spin rotators may be possible without requiring a realignment of snakes; this will be explored in the future. Another more expensive solution with a snake in the W-line has previously been discussed in Reference [1].

REFERENCES

1. I. Alekseev et al., *Design Manual Polarized Proton Collider at RHIC*, (1998). (http://www.rhichome.bnl.gov/RHIC/Spin/design)
2. T. Roser, "First Polarized Proton Collisions at RHIC", these proceedings (2003).
3. S. Y. Lee, *Spin Dynamics and Snakes in Synchrotrons*, World Scientific, Singapore (1997).
4. George Arfken, *Mathematical Methods for Physicists*, Academic Press, New York (1970).
5. W. MacKay and S. Peggs, "Accelerator Physics Coordinate Conventions", RHIC/AP/12 (1993).
6. M. J. Syphers, "Spin Motion through Helical Dipole Magnets", AGS/RHIC/SN No. 020, (1996).

Matching of Siberian Snakes

Georg H. Hoffstaetter

Cornell University, Ithaca/NY

Abstract. It is shown how one can choose suitable combinations of Siberian Snakes and betatron phase advances to optimize the ability of a ring to accelerate a polarized beam with little reduction of polarization. In an analysis of HERA-p, these methods have lead to a 14–fold increase of the vertical beam emittance for which polarization could be preserved. This is not only impressive as a result, but also the methods which lead to this result are very interesting. They contain detailed spin-orbit tracking and the application of the Froissart-Stora Formula for higher order spin-orbit resonances, for which an algorithm of determining resonance strength has been found.

INTRODUCTION

In order to produce polarized proton or deuteron beams at high energy, a polarized source is used and the beam is accelerated from low energy with as little loss of polarization as possible. Subsequently, the beam has to be stored at high energy with little depolarization for a long time, e.g. for many millions of turns around a storage ring.

While the polarized beam travels along the azimuth θ of the circular accelerator, its spin motion is influenced by external fields. Ideally, the center of the beam would travel in a horizontal plane and would only experience a vertical magnetic field so that a vertical spin would not precess and would therefore remain vertical from turn to turn. But even with field perturbations, a particle which travels along a periodic orbit and defines the center of a beam has a spin $\vec{n}_0$ which is periodic from turn to turn. However, $\vec{n}_0$ is in general not vertical. Other particles of the beam oscillate around the closed orbit during their motion around the circular accelerator and experience additional electromagnetic fields. These oscillations in phase space are described by $\vec{z}(\theta)$. The spins of those particles precess around a vector which differs from the vector around which particles on the closed orbit precess. The components of this vector which are perpendicular to $\vec{n}_0$ are expressed in the complex plane as $\omega(\vec{z}(\theta), \theta)$. These components cause a rotation of spins away from $\vec{n}_0$ and therefore cause depolarization of the beam. The depolarization is enhanced when the spin precession frequency around $\vec{n}_0$, the so called closed orbit spin tune ν_0, is in resonance with a Fourier component of $\omega(\vec{z}(\theta), \theta)$. The strength of this depolarizing effect is characterized by the first order resonance strength,

$$\varepsilon_\kappa = \lim_{N \to \infty} \frac{1}{2\pi N} \int_0^{2\pi N} \omega(\vec{z}(\theta), \theta) e^{-i\kappa\theta} d\theta . \tag{1}$$

Since the components of the phase space vector oscillate with the transverse tunes Q_x, Q_y, and the synchrotron tune Q_τ, resonances occur when the closed orbit spin tune ν_0 is a linear combination of these orbital tunes, i.e. $\nu_0 = j_0 + \vec{j} \cdot \vec{Q} = \kappa$ where all components

j_k of $\vec{j}$ are integers. Often the spin motion is approximated by the single resonance model (SRM) where only the dominant Fourier component of $\omega(\vec{z}(\theta), \theta)$ is taken into account and all other Fourier components are neglected.

The spin motion in the SRM is simple enough to be solved analytically. If the closed-orbit spin tune changes linearly, i.e. $dv_0/d\theta = \tilde{\alpha}$, and if the spin is initially choses as vertical, then the vertical spin component long after the resonance is crossed is given by the Froissart–Stora formula,

$$s_3(\infty) = 2e^{-\pi \frac{\varepsilon_K^2}{2\tilde{\alpha}}} - 1. \tag{2}$$

For slow crossing of the resonance, i.e. small $\tilde{\alpha}$, the spin is flipped from vertically up to vertically down and for very fast crossing of the resonance, the spin is hardly affected and remains vertically up.

The Froissart–Stora formula is regularly used to describe the reduction of polarization due to vertical betatron oscillations during resonance crossing in accelerators where the closed–orbit spin tune v_0 changes with energy. These descriptions were normally restricted to first order resonances, flat rings, and $v_0 = G\gamma$. In principle, also higher–order resonances can be treated in the SRM and could then be described by the Froissart–Stora formula. However, it has not been clear how to obtain the strength of higher order resonances since they cannot be computed by equation (1). Due to the nonlinear character of spin rotations, higher–order resonances can appear even when ω has only first order Fourier coefficients as in the case of linearized orbit motion. In fact, all higher order resonances which will be presented here have been computed with an $\omega(\vec{z}, \theta)$ which is linear in $\vec{z}$.

Since orbit tunes cannot be close to 0.5, Siberian Snakes which fix the closed orbit spin tune to 0.5 avoid first order resonances completely. Higher order resonances then become important, and in the following, it will be demonstrated how the influence of higher order resonances can be computed and minimized. It will even be shown that a Froissart–Stora like formula can be applied and therefore resonance strength can be computed quantitatively that characterize phenomena like spin flipping and depolarization at higher order resonances.

Motion of Spin Fields and the Invariant Spin Field

To achieve a more complete description of the depolarization process it is helpful to analyze not only the motion of individual spins but to investigate the dynamics of the spin field of a beam. A spin field $\vec{f}(\vec{z}, \theta)$ with $|\vec{f}| = 1$ describes the polarization of the beam by determining that each particle with the phase space coordinate $\vec{z}$ has the polarization direction $\vec{f}(\vec{z}, \theta)$ while it travels along the ring's azimuth θ. While each spin precesses around a vector $\vec{\Omega}(\vec{z}, \theta)$, the precession of the spin field is described by

$$\frac{d}{d\theta}\vec{f} = \partial_\theta \vec{f} + [\vec{v}(\vec{z}, \theta) \cdot \partial_{\vec{z}}]\vec{f} = \vec{\Omega}(\vec{z}, \theta) \times \vec{f}. \tag{3}$$

If all particles of a beam are initially completely polarized parallel to each other, the polarization state of the beam is in general not 2π-periodic and the beam polarization can change from turn to turn. A special spin field $\vec{n}(\vec{z}, \theta)$ which is 2π-periodic in θ is called an invariant spin field or Derbenev–Kondratenko $\vec{n}$-axis,

$$\vec{n}(\vec{z}, \theta + 2\pi) = \vec{n}(\vec{z}, \theta) . \quad (4)$$

If the spin of each particle in a beam is initially parallel to $\vec{n}(\vec{z}, \theta)$, particles get redistributed in phase space during one turn, but their spins will stay parallel to the invariant spin field. The spin field of the beam is then in an equilibrium state. Note that $\vec{n}(\vec{z})$ is usually not an eigenvector of the one turn spin transport matrix $\underline{R}(\vec{z})$ at some phase space point since the spin of a particle has changed after one turn around the ring, but the eigenvector would not have changed.

However, although it has been straightforward to define $\vec{n}(\vec{z}, \theta)$, it is not easy to calculate this spin field in general and much effort has been spent on this topic, mostly for electrons at energies up to 46 GeV. All algorithms developed before the polarized proton project at HERA-p [1, 2, 3, 4] rely on perturbation methods at some stage, and either do not go to high enough order [5, 6] or have problems with convergence at high order and high proton energies [7, 8, 9].

It can be shown that no spin field $\vec{f}(\vec{z}, \theta)$ has a time averaged polarization that is larger than that of the invariant spin field, which is given by the phase space average $P_{lim} = \langle \vec{n}(\vec{z}) \rangle_{\vec{z}}$. Furthermore, the invariant spin field is periodic from turn to turn and can therefore be used to define a basis to describe spin motion for particles which oscillate around the closed orbit due to their phase space amplitudes $\vec{J}$. The spin precessions in this coordinate system can be used to define an amplitude-dependent spin tune that gives rise to an amplitude dependent resonance condition, $v(\vec{J}) = j_0 + \vec{j} \cdot \vec{Q}$.

The Froissart–Stora Formula for Higher–Order Resonances

For the SRM the invariant spin field $\vec{n}$, the time average maximum polarization P_{lim}, and the amplitude-dependent spin tune $v(\vec{J})$ can be computed analytically, and are illustrated in Fig. 1. The spin tune is given by

$$v = \text{sig}(\delta)\sqrt{\delta^2 + \varepsilon_\kappa^2} + \kappa , \quad (5)$$

where κ is the spin tune at which the resonance condition is satisfied and $\delta = v_0 - \kappa$ is the distance of the closed orbit spin tune from the resonance. In this formula and in Fig. 1, it is apparent that the spin tune jumps by twice the resonance strength at the location where the closed orbit spin tune satisfies the resonance condition.

Before the investigations in [3], it was not known how resonance strengths of higher order could be computed since they cannot be computed by equation (1). But once one is able to compute the amplitude-dependent spin tune, there is however a simple method: compute the amplitude-dependent spin tune and observe its jump where the closed orbit spin tune satisfies a resonance condition. The jump is $2\varepsilon_\kappa$.

In [3], it has been shown that a SRM for the amplitude dependent spin tune close to an isolated resonance can be derived. However, here we simply track a beam that is initially

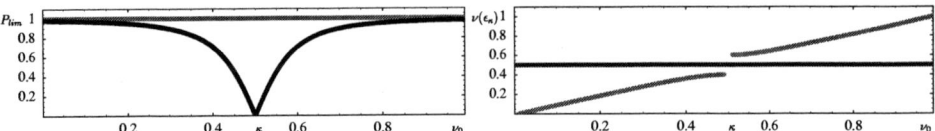

FIGURE 1. P_{lim} and the amplitude-dependent spin tune $v(\varepsilon_\kappa)$ for the SRM in the vicinity of $v_0 = \kappa$, for $\kappa = 0.5$ and $\varepsilon_\kappa = 0.1$.

polarized parallel to $\vec{n}(\vec{z})$ and accelerate it through a higher order resonance condition of the amplitude-dependent tune. If the remaining polarization in the direction of the new $\vec{n}$-axis at high energy follows the Froissart–Stora Formula, then we know that the strength of the higher order resonance has been determined.

In Fig. 2 (left), P_{lim} and v are shown after 4 Siberian Snakes were added to the simulation of the HERA-p optics used in 2002. P_{lim} is reduced at two resonances with $v = 2Q_y$.

The spins of a set of particles were set parallel to the invariant spin field $\vec{n}(\vec{z})$ so that all had $J_S = \vec{n}(\vec{z}) \cdot \vec{S} = 1$ at the momentum of 801 GeV/c. The $\vec{n}$-axis had been computed by stroboscopic averaging [1]. The beam was then accelerated to 804 GeV/c at various rates of acceleration. The average $\bar{J}_S$ over the tracked particles after acceleration is plotted versus rate of acceleration in Fig. 2 (right) together with the prediction of the Froissart–Stora formula for which the resonance strength ε_{2Q_y} has been determined from the tune jump. The parameter $\tilde{\alpha}$ is proportional to the energy increase per turn d_E and is determined from the tune slope $\frac{\Delta v}{\Delta E}$ in Fig. 2 (center) by the relation $\tilde{\alpha} = \frac{1}{2\pi} \frac{\Delta v}{\Delta E} d_E$. The polarization obtained by accelerating particles through the second order resonance agrees remarkably well with the Froissart–Stora formula. For the slow acceleration of about 50 keV per turn in HERA-p, the polarization would be completely reversed for particles with an amplitude of 0.75 sigma for the chosen scheme of four Siberian Snakes. This would lead to a net reduction of beam polarization, since the spins in the center of the beam are not reversed.

The two data points at the largest rate of acceleration are lower than predicted by the Froissart–Stora formula. One possible explanation is the following: at very large rates of acceleration, the resonance region is crossed so quickly that the spin motion is hardly disturbed. But when the invariant spin field $\vec{n}_-$ before the resonance region is not parallel to the invariant spin field $\vec{n}_+$ after the resonance region, then the spins which initially had $J_S = \vec{n}(\vec{z}) \cdot \vec{S} = 1$ will approximately have $J_S = \vec{n}_- \cdot \vec{n}_+$ after the resonance region is crossed. This is smaller than the Froissart–Stora prediction, which approaches 1 for large rate of acceleration.

SNAKE MATCHING

In order to minimize depolarizing resonance effects, firstly Siberian Snakes are inserted in the ring to fix the closed orbit spin to 0.5, which avoids first order resonances of the closed orbit spin tune. In Fig. 3 (left), the energy dependences of P_{lim} with and without

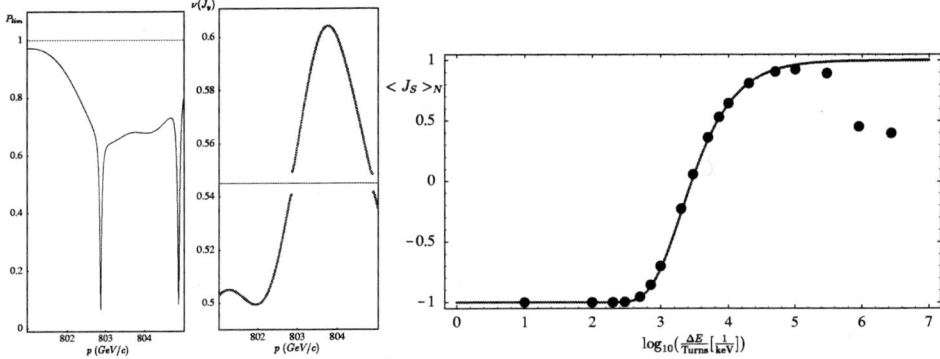

FIGURE 2. Left: P_{lim} and ν for a second–order resonance after an addition of a 4-snake scheme to the HERA-p optics that was used in 2002 with a 0.75σ vertical amplitude of 2.25π mm mrad. Right: The average $\langle J_S \rangle_N$ over N particles after acceleration from 801 GeV/c to 804 GeV/c with different rates of acceleration (*points*) and the prediction of the Froissart–Stora formula (*curve*) using parameters ε_{2Q_y} and $\tilde{\alpha}$ obtained from ν in the center plot.

Siberian Snakes are overlaid, which shows that not all resonance effects are removed by Siberian Snakes. In the energy regions where Siberian Snakes do not avoid the reduction of P_{lim}, the perturbations to spin motion in each FODO cell of HERA-p add up so that the resonance strengths are very large.

When designing a circular accelerator for polarized beams, one should therefore not only insert Siberian Snakes but also minimize the divergence or spread of the $\vec{n}$ – axis, and this maximizes P_{lim}. Since the spin tune jump at higher order resonances relates to their strength, one should additionally try to minimize these jumps. If possible one should avoid that the spin tune crosses a low order resonance line. Figure 3 (right) shows the amplitude–dependent spin tune.

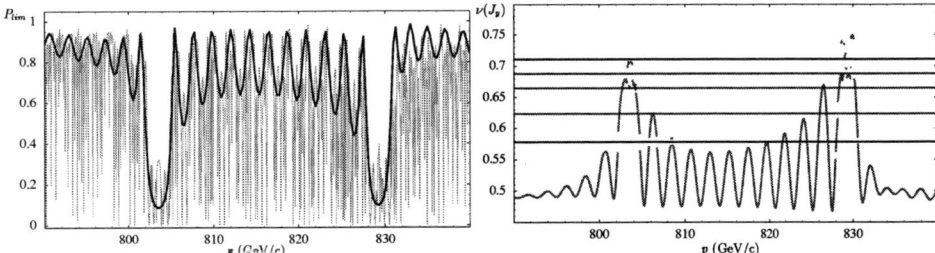

FIGURE 3. Left: First order P_{lim} with and without Siberian Snakes. Right: The amplitude–dependent but orbital–phase–independent spin tune ν for four Siberian Snakes in HERA-p. From top to bottom, the following resonance lines are drawn: $\nu = 1 - Q_y$, $\nu = 3 - 8Q_y$, $\nu = 5 - 15Q_y$, $\nu = 16Q_y - 4$, $\nu = 2Q_y$. The strength of these higher–order resonances can be deduced from the tune jumps.

Snake matching is the procedure proposed here for minimizing the spread of the $\vec{n}$ – axis by minimizing the spin orbit coupling integrals. At azimuth θ_0, the spin orbit coupling integrals are defined as

$$I_k^\pm = -i \int_{\theta_0}^{\theta_0+2\pi} \omega(\vec{v}_k^\pm, \theta) e^{i\{\pm Q_k(\theta-\theta_0) - \Psi\}} d\theta, \tag{6}$$

where $\vec{v}_k^\pm$ are the eigenvectors of the one turn beam transport matrix and ω describes the spin rotation as in equation (1) for linearized phase space motion. The spin phase advance from θ_0 to θ is Ψ, and $Q_k(\theta - \theta_0)$ is the orbit phase advance associated with the eigenmode $\vec{v}_k^\pm$. When the spin–orbit–coupling integrals for θ_0 are minimized, the opening angle of the invariant spin field at θ_0 for the approximation of linear spin–orbit motion is also minimized [3].

The vertical spin–orbit–coupling integrals from the first regular FODO cell to the last FODO cell of a regular arc in HERA-p will be denoted by $\hat{I}_y^+$ and $\hat{I}_y^-$ and the azimuths of the beginnings of the 4 regular arcs are θ_1, θ_2, θ_3, and θ_4 as shown in Fig. 4. The central points of the South, West, North, and East straight sections are denoted by S, W, N, and E. The spin phase advances between the arcs are suitably manipulated by the snake angles φ_E, φ_N, and φ_W. The closed orbit spin tune is then adjusted to 0.5 by φ_S. A snake angle ϕ, i.e. the angle that a Siberian Snake's horizontal rotation axis makes to the radial direction, is important here since a Siberian Snake not only flips the spin by 180°, but also causes a spin phase advance of 2ϕ around the vertical. The spin phase advance between θ_i and θ_j is denoted by Ψ_{ij}. Figure 4 (left) also shows eight flattening snakes, one in each straight section. They compensate the non-flat regions of HERA-p.

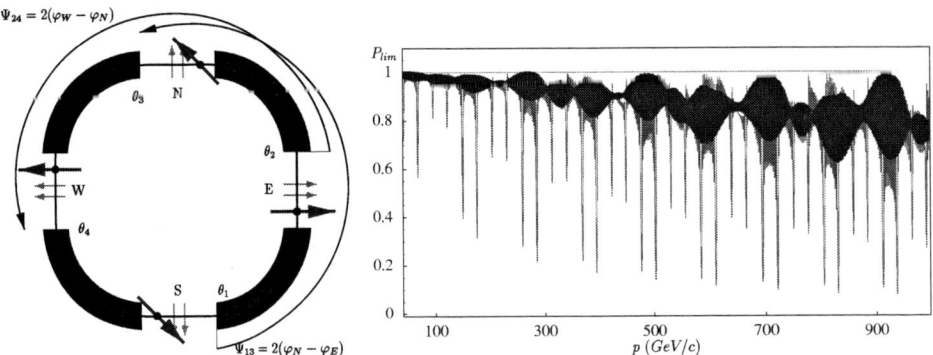

FIGURE 4. Left: The spin phase advance from the beginning of one regular arc to the beginning of the regular arc on the opposite side of the ring. **Right:** Improvement of linearized P_{lim} by matching 4 snake angles and the orbital phases. The snake arrangement is $(0\frac{\pi}{2}\frac{\pi}{2}\frac{\pi}{2})^*$ (*dark foreground curve*). As a comparison, P_{lim} from linearized spin–orbit motion is shown for the same HERA-p optics with a $(\frac{\pi}{4}0\frac{\pi}{4}0)$ snake scheme (*light background curve*).

With these 8 flattening snakes and one Siberian Snake in each of the straight sections, the spin phase advance from θ_1 to θ_3 is given by $\Psi_{13} = \Psi_{1E} - 2\varphi_E - \Psi_{EN} + 2\varphi_N + \Psi_{N3}$. Here, the fact that a Siberian Snake rotates the coordinate system for spins was used, so that the spin phase in the spin orbit coupling integral changes its sign. The terms due to Ψ cancel and the total spin phase advance is solely determined by the snake angles and is therefore independent of energy: $\Psi_{13} = 2(\varphi_N - \varphi_E)$ and $\Psi_{24} = 2(\varphi_W - \varphi_N)$. The orbital phase advance $\Phi_y(\theta_3) - \Phi_y(\theta_1)$ also does not depend on energy. For simplicity, $\Phi_y(\theta_j) - \Phi_y(\theta_i)$ will now be denoted by Φ_{ij}.

The spin–orbit–coupling integrals at the South interaction point contain the following contributions from the 4 regular arcs:

$$I^+_{arcs} = \hat{I}^+ e^{i(-\Psi_{S1}+\Phi_{S1})}(1+e^{i[2(\varphi_E-\varphi_N)+\Phi_{13}]}) \tag{7}$$
$$+(\hat{I}^-)^* e^{i(2\varphi_E-\Psi_{SE}+\Psi_{E2}+\Phi_{S2})}(1+e^{i[2(\varphi_W-\varphi_N)+\Phi_{24}]}),$$

$$I^-_{arcs} = \hat{I}^- e^{i(-\Psi_{S1}-\Phi_{S1})}(1+e^{i[2(\varphi_E-\varphi_N)-\Phi_{13}]}) \tag{8}$$
$$+(\hat{I}^+)^* e^{i(2\varphi_E-\Psi_{SE}+\Psi_{E2}-\Phi_{S2})}(1+e^{i[2(\varphi_W-\varphi_N)-\Phi_{24}]}).$$

This shows that it is always possible to cancel one of the spin–orbit coupling integrals by choosing the snake angles so that the spin perturbation produced in one of the arcs is canceled by the arc on the opposite side of the ring. Since $|\hat{I}^+|$ and $|\hat{I}^-|$ are different, neighboring arcs can in general not compensate each other.

To cancel both spin–orbit integrals in (8), 4 phase factors have to be -1. This requires

$$2(\varphi_E - \varphi_N) + \Phi_{13} = \pi \bmod 2\pi, \tag{9}$$
$$2(\varphi_E - \varphi_N) - \Phi_{13} = \pi \bmod 2\pi, \tag{10}$$
$$2(\varphi_W - \varphi_N) + \Phi_{24} = \pi \bmod 2\pi, \tag{11}$$
$$2(\varphi_W - \varphi_N) - \Phi_{24} = \pi \bmod 2\pi. \tag{12}$$

For arbitrary betatron phase advances, this equation cannot be solved by a choice of snake angles, since there are only two free parameters which contain the snake angles. However, the betatron phase advances can be changed appropriately. Subtraction of the first two equations leads to the requirement that the betatron phase advance from θ_1 half way around the ring to θ_3 be an odd or even multiple of π. The same is true for the phase advance from θ_2 to θ_4. Correspondingly, the spin phase advance over these regions has to be an odd multiple of π when the orbit phase advance is an even multiple and vice versa. With a rather benign change of the vertical optics in HERA-p which does not change the vertical tune, the contribution of the regular arcs to both spin–orbit–coupling integrals can thus be canceled when 4 Siberian Snakes are in HERA-p.

In the following, we characterize snake schemes by their snake angles starting in the South: e.g. $(\phi_S \phi_E \phi_N \phi_W)$. The snake scheme $(0 \frac{\pi}{2} \frac{\pi}{2} \frac{\pi}{2})^*$ has $\Psi_{13} = 0$ and $\Psi_{24} = 0$. The star indicates that the betatron phase advance has also been used for snake matching. For this snake scheme, the betatron phase advances from θ_1 to θ_3 and from θ_2 to θ_4 were adjusted to be odd multiples of π. The maximum time average polarization P_{lim} for linearized spin–orbit motion is plotted (dark foreground curve) in Fig. 4 (right) for the complete range of HERA-p momenta. As a comparison, P_{lim} for a standard snake scheme $(\frac{\pi}{4} 0 \frac{\pi}{4} 0)$ (light background curve) is also shown. The latter scheme and similar symmetric schemes were originally considered advantageous by a popular opinion [10], mostly due to their symmetry.

For linearized spin–orbit motion, the complete snake match of the arcs in HERA-p indeed eliminates all strong reductions of P_{lim} over the complete momentum range.

Schemes with 8 snakes:

Although eight snakes are not very practical for HERA-p, significant improvements are in principle possible when 8 snakes are used. A similar, only slightly more elaborate analysis shows that a choice of 8 snakes in the scheme characterized as $(\frac{\pi}{2}0000000)^*$ can cancel the contributions of the regular arc part of individual octants against each other if betatron phase advances are chosen appropriately. In this snake scheme, Φ_{12} was changed to $2\pi \times 8.5$ and Φ_{34} was changed to $2\pi \times 7.5$ without changing the vertical tune.

There is a second scheme with 8 snakes which also cancels the contribution of individual octants. This snake scheme is referred to as $(\frac{\pi}{2}abc0\text{-}c\text{-}b\text{-}a)^*$.

Nonlinear Spin Dynamics for Vertical Particle Motion

To check whether the improvements of spin motion, obtained in the framework of linearized spin–orbit motion, survive when higher–order effects are considered, P_{lim} and v has been calculated by the SODOM-2 algorithm with the code SPRINT [11]. The result for the Siberian Snake scheme $(\frac{\pi}{4}0\frac{\pi}{4}0)$, which was one of those that used to be considered advantageous by a popular opinion, is shown for the South interaction point of HERA-p in Fig. 5. Here many higher–order resonances are revealed, causing strong reduction of P_{lim} and there are corresponding strong variations of the amplitude-dependent spin tune v. The strongest spin tune jumps occur in the critical energy regions, mostly at the second order resonance $v = 2Q_y$, which is indicated by the top line [12].

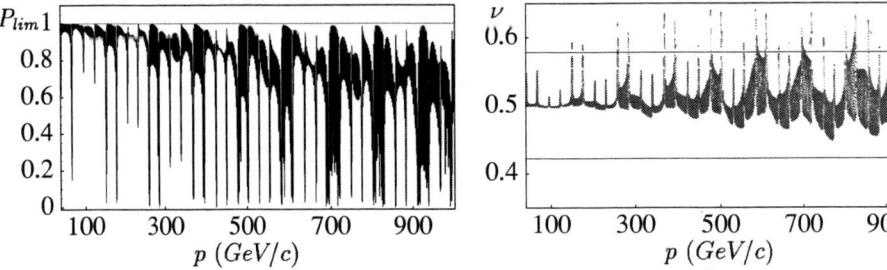

FIGURE 5. P_{lim} and v for particles with a 2.5σ vertical amplitude in the HERA-p lattice of the year 2002 with the standard scheme $(\frac{\pi}{4}0\frac{\pi}{4}0)$. The second–order resonances $v = 2Q_y$ and $v = 1 - 2Q_y$ are indicated by the two lines in the right graph.

P_{lim} and v for higher–order spin dynamics in the snake–matched and phase–advance–matched HERA-p with the snake scheme $(0\frac{\pi}{2}\frac{\pi}{2}\frac{\pi}{2})^*$ are shown in Fig. 6. While the overall behavior of P_{lim} over the complete acceleration range of HERA-p looks similar to the result obtained with linearized spin–orbit motion, which was displayed in Fig. 4 (right), higher–order effects become strong at high energies. But the spin tune spread at momenta below 400 GeV/c is small, and higher–order effects seem to be benign. The advantage over the snake scheme $(\frac{\pi}{4}0\frac{\pi}{4}0)$ becomes clear: v comes close to a second order resonance at much fewer places and only exhibit spin tune jumps which are much weaker than those shown in Fig. 5.

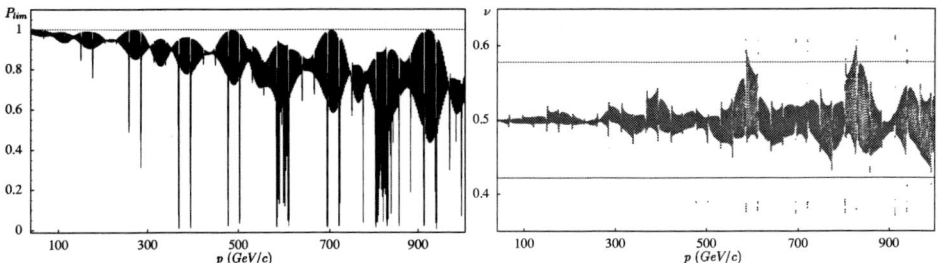

FIGURE 6. P_{lim} and ν for a 2.5σ vertical amplitude after the betatron phase advance between opposite regular arc structures was adjusted to be an odd multiple of π in the $(0\frac{\pi}{2}\frac{\pi}{2}\frac{\pi}{2})^*$ scheme. The resonances $\nu = 2Q_y$ and $\nu = 1 - 2Q_y$ are indicated by the two lines in the right graph.

Allowed Beam Sizes

After this optimization of snake schemes and the betatron phase advance, one finally has to check if, indeed, less depolarization occurs when particles are accelerated across the critical momentum region from 800 to 806 GeV/c with a typical rate of acceleration of 50 keV per turn. For that, one distributes particles in phase space and starts spins parallel to the invariant spin field $\vec{n}(\vec{z})$. After the acceleration, one then checks whether the spins are still parallel to the new invariant spin field at the increased energy. The average projection of the spins onto the new $\vec{n}$-axis, $\bar{J}_S = \langle \vec{n} \cdot \vec{S} \rangle$ is shown in Fig. 7 (left) for different vertical oscillation amplitudes and for different snake schemes.

For the standard snake scheme $(\frac{\pi}{4} 0 \frac{\pi}{4} 0)$, only the part of the beam with less than 1π mm mrad vertical amplitude can remain polarized. For the scheme $(\frac{3\pi}{4} \frac{3\pi}{8} \frac{3\pi}{8} \frac{\pi}{4})$, which had been found by trial and error, phase space amplitudes up to 4π mm mrad are allowed. Finally, the snake matched scheme $(0\frac{\pi}{2}\frac{\pi}{2}\frac{\pi}{2})^*$ gives the most stable spin motion and Fig. 7 (left) shows that vertical amplitudes of up to 8π mm mrad are allowed.

The two different snake matches for 8-snake schemes have also been tested. In Fig. 7 (right), it is shown that both lead to very small spin tune variations even compared to the matched 4-snake scheme. As shown in Fig. 7 (left), the more effective of the two 8-snake schemes stabilizes spin motion up to a vertical amplitude of 14π mm mrad.

These results show that it is not possible to give a simple formula for the number of snakes which are required for a given accelerator since different snake schemes with the same number of snakes lead to very different stability of spin motion since the required number of snakes depends strongly on the chosen snake angles.

ACKNOWLEDGMENTS

A collaboration with Desmond Barber, Klaus Heinemann, Mathias Vogt, and Kaoru Yokoya has been very valuable in establishing the results presented here. Associating the gaps of the amplitude–dependent spin tune with resonance strength was inspired in 1999 by a comment of A. Lehrach during a talk by M. Vogt.

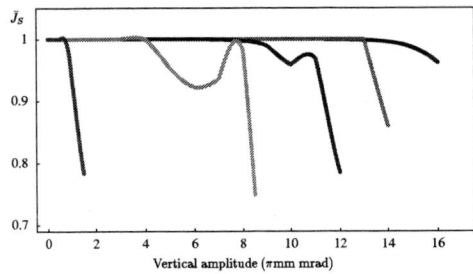
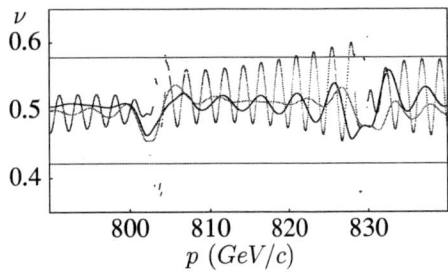

FIGURE 7. Left: The average spin action $\bar{J}_S$ at 806 GeV for particles which started initially with $J_S = \vec{n} \cdot \vec{S} = 1$ at 800 GeV/c. From the leftmost to the rightmost curve: 1) The standard scheme which stabilizes spin motion for particles within 1π mm mrad, 2) a 4-snake scheme found by trial and error which stabilizes within 4π mm mrad, 3) the matched 4-snake scheme which stabilizes within 8π mm mrad, 4) the matched 8-snake scheme which stabilizes within 13π mm mrad, 5) and the matched 8-snake scheme which stabilizes within 14π mm mrad. **Right:** Improvement of $\nu(J_y)$ by matching 8 snake angles and the orbital phases. The snake arrangements are $(\frac{\pi}{2}0000000)^*$ (*dark*), $(\frac{\pi}{2}abc0\text{-}c\text{-}b\text{-}a)^*$ (*light*), and $(0\frac{\pi}{2}\frac{\pi}{2}\frac{\pi}{2})^*$ (*curve with largest oscillations*). The resonances $\nu = 2Q_y$ and $\nu = 1 - 2Q_y$ are indicated by lines.

REFERENCES

1. K. Heinemann and G. H. Hoffstaetter, A tracking algorithm for the stable spin polarization field in storage rings using stroboscopic averaging, *Phys. Rev.* **E 54**, 4240–4255 (1996).
2. G. H. Hoffstaetter, M. Vogt, and D. P. Barber, Higher–order effects in polarized proton dynamics, *Phys. Rev.* **ST–AB, 2(11)**, 1–15 (1999).
3. G. H. Hoffstaetter, Aspects of the invariant spin field for high energy polarized proton beams, habilitation, *Darmstadt Univ. of Tech.* (January 2000).
4. M. Vogt, Bounds on the maximum attainable equilibrium spin polarization of protons in HERA, Dissertation, *Universität Hamburg*, DESY–THESIS–2000–054 (December 2000).
5. A. W. Chao, Evaluation of radiative spin polarization in an electron storage ring, *Nucl. Instrum. Methods* **180**, 29 (1981).
6. Yu. Eidelman and V. Yakimenko, The application of Lie methods to the spin motion in nonlinear collider fields, *Particle Accelerators*, **45**, 17–35 (1994).
7. S. R. Mane, Electron-spin polarization in high–energy storage rings II. Evaluation of the equilibrium polarization, *Phys. Rev.* **A(36)**, 120–130 (1987).
8. K. Yokoya, Non–perturbative calculation of equilibrium polarization of stored electron beams, *Report KEK–92–6*, Tsukuba (1992).
9. V. Balandin and N. Golubeva, Nonlinear spin dynamics, *Proceedings of the XV International Conference on High–energy Particle Accelerators*, Hamburg, pp. 998–1000 (1992).
10. V. Ptitsin and Yu. M. Shatunov, The investigation of spin resonances in an accelerator with Siberian Snakes, in C. W. de Jager, T. J. Ketel, P. J. Mulders, J. E. J. Oberski, and M. Oskam-Tamoezer, editors, *Proceedings SPIN96*, pp. 516–518, World Scientific (1996).
11. K. Yokoya, An algorithm for calculating the spin tune in circular accelerators, *Report DESY–99–006* (1999).
12. D. P. Barber, G. H. Hoffstaetter, and M. Vogt, Using the amplitude–dependent spin tune to study high order spin–orbit resonances in storage rings, *AIP Conf. Proc. of the 14th intl. Spin Physics Symposium, SPIN2000*, **570**, pp. 751–755 (2001).

Workshop Highlights and Summary[1]

Thomas Roser

Brookhaven National Laboratory, Upton, NY 11973, USA

INTRODUCTION

The acceleration of polarized proton beams with energies in the range of 3 - 30 GeV in circular accelerators is very challenging since depolarizing resonances are strong enough to cause significant if not complete depolarization but the beam energy is too low for practical designs of full Siberian snakes. Over the last thirty years, however, several techniques have been developed to overcome depolarizing resonances with significant success.

During this workshop the experience and new methods of accelerating polarized beams in circular accelerators were presented and discussed with the goal of planning a strategy to raise the beam polarization in the Brookhaven Alternating Gradient Synchrotron (AGS) to the highest possible level. The AGS serves as injector to the 500 GeV polarized proton collider RHIC. The planned two-spin experiments at RHIC are very sensitive to the level of polarization both for improved experimental statistics and reduced systematic errors.

For low-energy, rapid-cycling machines such as the AGS, only first-order depolarizing resonances are important. The resonance condition for the spin tune v_{sp} is:

$$v_{sp} = G\gamma = n \pm m v_y \pm k v_x \quad (m, k = \{0, 1\})$$

There are two main types of first order depolarizing spin resonances: imperfection resonances, which are driven by magnet errors and misalignments, and intrinsic resonances, driven by the focusing fields. The strengths of both types of resonances increases with beam energy. The intrinsic resonances themselves can be grouped into three categories: strong intrinsic resonances that lead to complete depolarization or partial spin flip; weak intrinsic resonances that cause some depolarization; and coupling resonances that are driven by the horizontal betatron motion coupled into the vertical plane by skew quadrupoles or solenoids.

During acceleration in the AGS to RHIC injection energy 42 imperfection resonances, 4 strong and 3 weak intrinsic resonances, and 4 coupling resonances are crossed. Each resonance can cause significant beam depolarization. The level of expected depolariza-

---

[1] Work performed under the auspices of the U.S. Department of Energy

tion or spin flip is given by the Froissart-Stora formula:

$$P_f/P_i = 2e^{-\frac{\pi|\varepsilon|^2}{2\alpha}} - 1,$$

where P_i and P_f are the polarizations before and after the resonance crossing, respectively, ε is the resonance strength, and α is the ramping speed. If the beam is accelerated through the resonance slowly or the resonance is very strong ($\alpha \ll |\varepsilon|^2$), the spin vector will adiabatically follow the stable spin direction resulting in complete spin flip. For fast acceleration or weak resonances ($\alpha \gg |\varepsilon|^2$)the polarization remains unchanged. However, for intermediate conditions partial depolarization or partial spin flip will occur.

The methods developed to overcome depolarization can then be grouped into "adiabatic methods", where the resonance strength is increased or the ramp rate decreased to achieve full spin-flip, and "non-adiabatic methods" where the resonance strength is reduced or the ramp rate increased until no polarization is lost. Examples of the former are the use of a localized spin rotator or 'partial Siberian snake' to make all the imperfection resonance strengths large and the use of a vertical rf dipole magnet to create a strong artificial spin resonance, overpowering the effect of the intrinsic resonances. The "non-adiabatic methods" were used early on and consist of the use of correction dipoles to minimize the strength of imperfection resonances and pulsed quadrupoles to jump the betatron tune during intrinsic resonance crossing, which increases the effective ramping speed.

EXPERIENCE AT EXISTING FACILITIES

Table 1 shows a compilation of the methods used at the three machines represented at this workshop. Also listed are the type of polarimeters used to measure the circulating beam polarization and what information was used to set-up the machine. Typically the beam polarization is measured at the end of the acceleration cycle. Using this information for machine set-up can be quite time-consuming and also makes it very difficult to diagnose depolarization from crossing a resonance during acceleration. Polarization measurement during the acceleration ramp makes this easier but typically takes a long time and also causes emittance growth that can itself lead to depolarization. The adiabatic methods allow at least the initial set-up to be performed using beam information which is easily available during the acceleration ramp.

Andreas Lehrach summarized the experience from the Cooler Synchrotron (COSY) in Juelich, Germany. Pulsed quadrupoles are used to overcome the intrinsic resonances and harmonic orbit correction dipoles are exited enough to cause full spin flip. Using the corrector dipoles to enhance instead of correcting the imperfection resonance strength requires a large aperture but is operationally more stable. The maximum energy reached is $G\gamma \approx 7$ with about 75 % polarization. The performance is presently limited due the marginal size of the tune jump for the strongest intrinsic resonances. It was also successfully demonstrated that proper modifications to the lattice (spin matching) can suppress the strength of individual weak intrinsic resonances. This method works well if the machine lattice has sufficient built-in flexibility.

The KEK Proton Synchrotron accelerated polarized protons during the 1980's and reached about 40% polarization at $G\gamma = 8.4$ and 25% at $G\gamma = 11.2$, as reported by Chihiro Ohmori. KEK used correction dipoles and pulsed quadrupoles to overcome intrinsic and imperfection resonances. The experience also showed that the tune jump was too small for the very strong intrinsic resonances of the KEK PS.

Leif Ahrens summarized the first experience of accelerating polarized protons at the Brookhaven AGS. During this first effort the non-adiabatic methods that were later used at KEK and COSY were developed. The 94 corrector dipoles were successfully used to overcome imperfection resonances although operationally the set-up and maintenance of the correct tuning of these many dipoles was quite time consuming. A very powerful, pulsed quadrupole system worked very well to overcome the intrinsic resonances and, after careful alignment, did not substantially increase the beam emittance. The strength of both the corrector dipoles and the tune jump system became marginal at the highest AGS energies. The maximum polarization reached was about 45% at $G\gamma = 41.5$.

Haixin Huang and Mei Bai reported on the recent efforts to accelerate polarized protons in the AGS for injection into RHIC. A partial Siberian snake is used to enhance the strength of all imperfection resonances and a vertical rf dipole magnet creates a strong artificial spin resonance by driving large coherent betatron oscillations, overpowering the effect of the intrinsic resonances. Both of these methods are adiabatic which makes them operationally more stable and also allows the set-up to be performed using beam information instead of the polarization information which takes a long time to acquire. Also both the partial Siberian snake and the rf dipole method are able to deal with very strong resonances. However, the three weak intrinsic resonances and the four weak coupling resonances generated by the partial snake solenoid cannot be treated by the vertical rf dipole method. Nevertheless, a polarization of about 40% was reached at the RHIC injection energy of $G\gamma = 46.5$.

ADDITIONAL METHODS AND PLANS

An alternative method, described by Haixin Huang, was recently tested at the AGS. With a strong enough partial Siberian snake both imperfection and intrinsic resonances can be overcome. The minimum strength is about 20% of a full snake and it also requires that the betatron tune is brought very close to an integer value which makes machine operation less stable. This method is analogous to the operation with a full Siberian snake with the main advantage that with a single device depolarization from all types of resonances can be avoided.

Based on the presented experience and discussions preferred methods for each type of depolarizing resonance are summarized in Table 1 under Plan 1. Plan 2 consists of a strong 25% partial Siberian snake that can cope with all resonances. Imperfection resonances are well treated with a partial Siberian snake. Coupling resonances are best avoided by building a partial snake that does not introduce orbit coupling. This can be accomplished using a helical dipole instead of a solenoid. Using the super-conducting magnet technology developed for the RHIC Siberian snakes a 20 - 30% partial snake can be build for the AGS. The same design produces a 5% partial snake if built as a warm

TABLE 1. Methods to overcome depolarizing spin resonances

	Imperfection resonances	Strong intrinsic resonances	Weak intrinsic resonances	Coupl. intr. resonances	Set-up information	Polarimeter	Meas. duration
COSY	corr. dipoles	pulsed quadrupoles	pulsed quadrupoles	N/A	Pol.	pC quasi-elast.	5 min.
KEK PS	corr. dipoles	pulsed quadrupoles	pulsed quadrupoles	N/A	Pol.	pC & pp elastic	2 min.
AGS (1983-1988)	94 corr. dipoles	10 puls. quadrupoles	10 puls. quadrupoles	N/A	Pol.	pC & pp elastic	10 min.
AGS (1994-2002)	5% solenoid snake	vertical rf dipole	Did nothing	Did nothing	Beam/Pol.	pC & pp elastic	10 min.
Plan 1	5% helical snake	vertical rf dipole	Intr. spin matching pulsed quadrupoles	N/A	Beam/Pol.	pC CNI	5 sec.
Plan 2	25% helical snake	25% helical snake	25% helical snake	N/A	Beam	pC CNI	5 sec.

magnet.

For weak intrinsic resonances there are several successful methods: a tune jump using pulsed quadrupoles has been demonstrated to work well, spin matching also works if the lattice is flexible enough, and a strong partial snake is expected to work as well. For strong intrinsic resonances, only the vertical rf dipole method has been demonstrated to be able to avoid depolarization although the accuracy of the experimental data still allows for a residual depolarization of a few percent. Proper set-up of the rf dipole requires very accurate tune control and a very small betatron tune spread, which is particularly difficult at higher beam intensities. For the strong intrinsic resonances the strong partial Siberian snake should eventually provide a better and operationally more stable solution.

SUMMARY

Based on this workshop a plan for upgrading polarized proton acceleration in the AGS was developed. The construction of a strong partial Siberian snake was initiated. Although in principle this single device would avoid all sources of depolarization in the AGS its construction, installation and commissioning will take several years. Also mismatch of the polarization direction at injection into the AGS will cause some depolarization. Plan 1 outlined above will be pursued in the meantime. A warm helical partial Siberian snake will replace the present solenoid snake. It will avoid the coupling resonances and can also be used in the future to avoid injection mismatch with the strong partial snake. Existing quadrupoles will be moved to locations where they can be used to suppress the weak intrinsic resonances as discussed at this workshop by Andreas Lehrach. This approach should give maximum polarization from the AGS as soon as possible and also provide a long term solution that is operationally simple and offers additional polarization improvements if the rf dipole method shows residual depolarization.

Participants List

Dr. L. A. Ahrens
Collider Accelerator Department
Brookhaven National Laboratory
Upton, NY 11973-5000 USA
Telephone: 631-344-4568
Fax: 631-344-5954
E-mail: ahrens@bnl.gov

Dr. B. B. Blinov
Department of Physics
University of Michigan
Ann Arbor, MI 48109-1120 USA
Telephone: 734-647-9032
Fax: 734-763-9694
E-mail: bblinov@umich.edu

Dr. M. Bai
Collider Accelerator Department
Brookhaven National Laboratory
Upton, NY 11973-5000 USA
Telephone: 631-344-3397
Fax: 631-344-5954
E-mail: mbai@bnl.gov

Dr. G. M. Bunce
Physics Department
Brookhaven National Laboratory
Upton, NY 11973-5000 USA
Telephone: 631-344-4771
Fax: 631-344-2562
E-mail: bunce@bnl.gov

Professor E. D. Courant
Collider Accelerator Department
Brookhaven National Laboratory
Upton, NY 11973-5000 USA
Telephone: 631-344-4609
Fax: 631-344-5954
E-mail: courant@bnl.gov

Professor Ya. S. Derbenev
Jefferson Memorial Lab
Newport News, VA 23606 USA
Telephone: 757-269-5051
Fax: 757-269-5024
E-mail: derbenev@jlab.org

Dr. H. Huang
Collider Accelerator Department
Brookhaven National Laboratory
Upton, NY 11973-5000 USA
Telephone: 631-344-5446
Fax: 631-344-5954
E-mail: huanghai@bnl.gov

Professor G. Hoffstaetter
Newman Laboratory
Cornell University
Ithaca, NY 14853-5001 USA
Telephone: 607-255-5197
Fax: 607-254-4552
E-mail: hoff@lns.cornell.edu

Professor A. D. Krisch
Spin Physics Center
University of Michigan
Ann Arbor, MI 48109-1120 USA
Telephone: 734-936-1027
Fax: 734-936-0794
E-mail: krisch@umich.edu

Dr. A. Lehrach
Forschungszentrum Jülich
Institut für Kernphysik, Postfach 1913
D-52425 Jülich GERMANY
Telephone: 49-2461-61-6453
Fax: 49-2461-61-2356
E-mail: a.lehrach@fz-juelich.de

Dr. A. M. T. Lin
Spin Physics Center
University of Michigan
Ann Arbor, MI 48109-2036 USA
Telephone: 734-763-9033
Fax: 734-763-9027
E-mail: alilin@umich.edu

Dr. W. W. MacKay
Collider Accelerator Department
Brookhaven National Laboratory
Upton, NY 11973-5000 USA
Telephone: 631-344-3076
Fax: 631-344-5954
E-mail: mackay@bnl.gov

Mr. V. S. Morozov
Spin Physics Center
University of Michigan
Ann Arbor, MI 48109-2036 USA
Telephone: 734-763-8161
Fax: 734-763-9027
E-mail: morozov@umich.edu

Dr. C. Ohmori
KEK-Accelerator Lab
1-1 Oho, Tsukuba-shi
Ibaraki-ken 305-0801 JAPAN
Telephone: 81-298-79-6119
Fax: 81-298-79-6130
E-mail: chihiro.ohmori@kek.jp

Professor Yu. F. Orlov
Lab for Elementary Particle Physics
Cornell University
Ithaca, NY 14853-5001 USA
Telephone: 607-255-3502
Fax: 607-254-4552
E-mail: orlov@lns.cornell.edu

Dr. T. Roser
Collider Accelerator Department
Brookhaven National Laboratory
Upton, NY 11973-5000 USA
Telephone: 631-344-7084
Fax: 631-344-5954
E-mail: roser@bnl.gov

Professor D. W. Sivers
Portland Physics Institute/UMich
4730 SW Macadam Avenue
Portland, OR 97201 USA
Telephone: 503-223-2680
Fax: 503-223-2750
E-mail: densivers@sivers.com

Dr. L. C. Teng
Argonne National Laboratory
Argonne, IL 60439 USA
Telephone: 630-252-3405
Fax: 630-252-4732
E-mail: teng@aps.anl.gov

Dr. W. B. Tippens
Division of Nuclear Physics /SC-23
U. S. Department of Energy
Washington, D.C. 20585-1290 USA
Telephone: 301-903-3904
Fax: 301-903-3833
E-mail: brad.tippens@science.doe.gov

Professor V. K. Wong
Spin Physics Center
University of Michigan
Ann Arbor, MI 48109-1120 USA
Telephone: 734-936-1027
Fax: 734-936-0794
E-mail: vkw@umich.edu

Dr. K. Yonehara
Spin Physics Center
University of Michigan
Ann Arbor, MI 48109-2036 USA
Telephone: 734-764-5114
Fax: 734-763-9027
E-mail: yonehara@umich.edu

Professor A. N. Zelenski
Collider Accelerator Department
Brookhaven National Laboratory
Upton, NY 11973-5000
Telephone: 631-344-8387
Fax: 631-344-5676
E-mail: zelenski@bnl.gov

Final Agenda
November 6-9, 2002 Workshop on Increasing the AGS Polarization

Wednesday, November 6, 2002 (Courtyard by Marriott) 19:00-21:00 Registration & Welcome Reception
Thursday, November 7, 2002 (Spin Physics Center, 1239 Kipke Drive)
Chair   Ya.S. Derbenev, J-Lab
09:00   Welcome and Introduction (15 min)   A.D. Krisch   Michigan
09:20   First Polarized Proton Collisions at RHIC (60 min)   T. Roser   Brookhaven
10:30   Coffee
11:00   AGS Beam Optics with a 20% Siberian snake (25 min)   E.D. Courant   Brookhaven/UM
11:25   Discussion (35 min)   ADK - leader
12:00   Thoughts & "Facts" from the AGS Polarized Proton
           Beam Runs during the 1980's (25 min)   L.A. Ahrens   Brookhaven
12:25   Discussion (35 min)   TR - leader
13:00   Lunch
Chair   L.C. Teng, Argonne
14:30   RF Dipole for Strong Depolarizing Resonances (25 min)   M. Bai   Brookhaven
14:55   Discussion (35 min)   ADK – leader
15:30   Overcoming Depol. Resonances At COSY (25 min)   A. Lehrach   COSY
15:55   Discussion (35 min)   TR – leader
16:30   Coffee
17:30   20% Partial Siberian Snake in the AGS (25 min)   H. Huang   Brookhaven
17:55   Discussion (35 min)   ADK – leader
Friday, November 8, 2002 (Spin Physics Center, 1239 Kipke Drive)
Chair   Yu. F. Orlov, Cornell
09:00   KEK Polarized Proton Beam Acceleration in the 1980's (25 min)   C. Ohmori   KEK/JHF
09:25   Discussion (35 min)   TR – leader
10:00   AGS Pulsed Quadrupoles History and Future (15 min)   A.D. Krisch   Michigan
10:15   Rapid 5% AGS Helical Snake (15 min)   T. Roser   Brookhaven
10:30   Discussion (30 min)   LCT – leader
11:00   Coffee and Tour
Chair   B.B. Blinov, Michigan
11:45   Pulsed Quads to Minimize Weak Intrin Res Depol (25 min)   M. Bai   Brookhaven
12:10   Discussion (35 min)   TR – leader
12:45   OPPIS Upgrade for 2003 Polarized Run in RHIC   A.N. Zelenski   Brookhaven
13:10   Discussion (35 min)   ADK – leader
14:00   Lunch at China Gate Restaurant (Main Campus)
16:00   Spin Physics Seminar in room 335 West Hall (Main Campus)
           "Deuteron Spin Flipping and Quantum Mechanics"   D.W. Sivers   Portland Phys Inst/UM
Saturday, November 9, 2002 (Spin Physics Center, 1239 Kipke Drive)
Chair   V.K. Wong, Michigan
09:00   AGS Lattice Changes to Eliminate Weak Intrinsic Res (25 min)   A. Lehrach   COSY
09:25   Discussion (35 min)   TR - leader
10:00   Cross Strong Intrin & Coup Res with Hor RF Dipole (25 min)   M. Bai   Brookhaven
10:25   Discussion (35 min)   LCT - leader
11:00   Coffee
11:30   The AGS CNI Polarimeter (25 min)   G.M. Bunce   Brookhaven
11:55   Discussion (35 min)   TR – leader
12:30   Roundtable: D.W. Sivers (15 min) Yu.F. Orlov (10 min)   ADK – leader
           C. Ohmori (10 min) Ya.S. Derbenev (10 min)
13:30   Lunch
Chair   A.N. Zelenski, Brookhaven
15:00   Spin Matching from AGS to RHIC (25 min)   W. MacKay   Brookhaven
15:25   Discussion (35 min)   LCT – leader
16:00   Matching of Siberian Snakes (25 min)   G. Hoffstaetter   Cornell
16:25   Discussion   ADK – leader
17:00   Coffee
Chair   C. Ohmori, KEK/JHF
17:30   Workshop Highlights and Summary   T. Roser   Brookhaven
20:00   Banquet   *Bella Ciao*

AUTHOR INDEX

A

Ahrens, L., 1, 9, 15, 40
Alekseev, I., 1
Alekseev, I. G., 77
Alessi, J., 1, 61

B

Bai, M., 1, 15, 40
Bechstedt, U., 30
Beebe-Wang, J., 1
Bravar, A., 77
Brennan, J. M., 1
Briscoe, B., 61
Brown, K. A., 1, 40
Bunce, G., 1, 77

C

Cameron, P., 1
Courant, E. D., 1

D

Deshpande, A., 1
Dhawan, S., 77
Dietrich, J., 30
Drees, A., 1

F

Fischer, W., 1
Fliller, III, R., 1

G

Gebel, R., 30
Glenn, W., 1, 40

H

Hiramatsu, S., 50
Hoffstaetter, G. H., 93
Huang, H., 1, 40, 77
Hughes, V., 77

I

Igo, G., 77

J

Jinnouchi, O., 1, 77

K

Kanavets, V., 77
Klenov, V., 61
Kokhanovski, S., 61
Kponou, A., 61
Krisch, A. D., 58
Krueger, K., 1
Kurita, K., 1, 77

L

Lehrach, A., 30, 67
Li, Z., 77
LoDestro, V., 61
Lorentz, B., 30
Lozowski, W., 77
Luccio, A. U., 1, 40

M

MacKay, W. W., 1, 40, 77, 84
Maier, R., 30
Makdisi, Y., 1, 77
Montag, C., 1, 40

O

Ohmori, C., 50
Okamura, M., 1

P

Pilat, F., 1
Prasuhn, D., 30
Ptitsyn, V., 1, 40

R

Ranjbar, V., 1, 40
Ranjbar, V. H., 67
Rescia, S., 77
Ritter, J., 61
Roser, T., 1, 9, 15, 40, 77, 103

S

Saito, N., 1
Sato, H., 50
Satogata, T., 1
Schnase, A., 30
Schneider, H., 30
Sivers, D. W., 81
Spinka, H., 1, 40
Stassen, R., 30
Stockhorst, H., 30
Svirida, D., 1
Svirida, D. N., 77
Syphers, M., 1

T

Tepikian, S., 1
Tojo, J., 1
Tölle, R., 30
Toyama, T., 50
Trbojevic, D., 1
Tsoupas, N., 1, 40, 84

U

Underwood, D., 1, 40

V

van Zeijts, J., 1

W

Whitten, C., 77
Wood, J., 77

Z

Zelenski, A., 1, 61
Zeno, K., 1, 40
Zubets, V., 61